GRADES 7-8

the Super Source®

Measurement

Cuisenaire Company of America, Inc.
White Plains, New York

Cuisenaire extends its warmest thanks to the many teachers and students across the country who helped ensure the success of *the Super Source®* series by participating in the outlining, writing, and field testing of the materials.

Managing Editors: Catherine Anderson and Alan MacDonell
Project Editor: Toni-Ann Guadagnoli
Contributing Writers: Deborah J. Slade and Carol Desoe
Production/Manufacturing Director: Janet Yearian
Production/Manufacturing Coordinator: Joan Lee

Design Director: Phyllis Aycock
Cover Design: Phyllis Aycock
Line Art and Composition: Eileen Sullivan

This book is published by Cuisenaire® Company of America, an imprint of Addison Wesley Longman, Inc.

Cuisenaire Company of America, Inc.
10 Bank Street
White Plains, New York 10602-5026
Customer Service: 800-237-3142

Printed in the United States of America
ISBN 1-57452-172-1
CCA015185

2 3 4 5 - SCG - 02 01 00 99

the Super Source®

Table of Contents

Using the Super Source®

The Super Source® Grades 7-8 continues the Grades K-6 series of activities using manipulatives. Driving ***the Super Source*** is Cuisenaire's conviction that children construct their own understandings through rich, hands-on mathematical experiences. There is a substantial history of manipulative use in the primary grades, but it is only in the past ten years that educators have come to agree that manipulatives play an important part in middle grade learning as well.

Unlike the K-6 series, in which each book is dedicated to a particular manipulative, the Grades 7-8 series is organized according to five curriculum strands: Number, Geometry, Measurement, Patterns/Functions, and Probability/Statistics. The series includes a separate book for each strand, each book containing activities in which students use a variety of manipulatives: Pattern Blocks, Geoboards, Cuisenaire® Rods, Snap™ Cubes, Color Tiles, and Tangrams.

Each book contains eighteen lessons grouped into clusters of 3–5 lessons each. Each cluster of lessons is introduced by a page of comments about how and when the activities within each lesson might be integrated into the curriculum.

The lessons in ***the Super Source*** follow a basic structure consistent with the vision of mathematics teaching described in the *Curriculum and Evaluation Standards for School Mathematics* published by the National Council of Teachers of Mathematics. All of the activities involve Problem Solving, Communication, Reasoning, and Mathematical Connections—the first four NCTM Standards.

HOW THE LESSONS ARE ORGANIZED

At the beginning of each lesson, you will find, to the right of the title, a list of the topics that students will be working with. GETTING READY, offers an *Overview,* which states what children will be doing, and why, and provides a list of *What You'll Need.* Specific numbers of manipulative pieces are suggested on this list but can be adjusted as the needs of your particular situation dictate. In preparation for an activity, pieces can be counted out and placed in containers or self-sealing plastic bags for easy distribution. Blackline masters that are provided for your convenience at the back of the book are also referenced on this materials list, as are activity masters for each lesson.

The second section, THE ACTIVITY, contains the heart of the lesson: a two-part *On Their Own,* in which rich problems stimulate many different problem-solving approaches and lead to a variety of solutions. These hands-on explorations have the potential for bringing students to new mathematical ideas and deepening skills. They are intended as stand-alone activities for students to explore with a partner or in a small group. *On Their Own* Part 2 extends the mathematical ideas explored in Part 1.

After each part of *On Their Own* is a *Thinking and Sharing* section that includes prompts you can use to promote discussion. These are questions that encourage students to describe what they notice, tell how they found their results, and give the reasoning behind their conclusions. In this way, students learn to verify their own results rather than relying on the teacher to determine if an answer is "right" or "wrong." When students compare and discuss results, they are often able to refine their thinking and confirm ideas that were only speculations during their work on the *On Their Own* activities.

The last part of THE ACTIVITY is *For Their Portfolio,* an opportunity for the individual student to put together what he or she has learned from the activity and discussion. This might be a piece of writing in which the student communicates results to a third person; it could be a drawing or plan synthesizing what has occurred; or it could be a paragraph in which the student applies the ideas from the activity to another area. In any case, the work students produce *For Their Portfolio* is a reflection of what they've taken from the activity and made their own.

The third and final section of the lesson is TEACHER TALK. Here, in *Where's the Mathematics?*, you as the teacher can gain insight into the mathematics underlying the activity and discover some of the strategies students are apt to use as they work. Solutions are also given, when such are necessary and/or helpful. This section will be especially helpful to you in focusing the *Thinking and Sharing* discussion.

USING THE ACTIVITIES

The Super Source is designed to fit into a variety of classroom environments. These can range from a completely manipulative-based classroom to one in which manipulatives are just beginning to play a part. You may choose to have the whole class work on one particular activity, or you may want to set different groups to work on two or three related activities. This latter approach does not require full classroom sets of a particular manipulative.

If students are comfortable working independently, you might want to set up a "menu"—that is, set out a number of activities from which students can choose. If this is the route you take, you may find it easiest to group the lessons as they are organized in the book—in small clusters of related activities that stimulate similar questions and discussion.

However you choose to use ***the Super Source*** activities, it would be wise to allow time for several groups or the entire class to share their experiences. The dynamics of this type of interaction, where students share not only solutions and strategies but also thoughts and intuitions, is the basis of continued mathematical growth. It allows students who are beginning to form a mathematical structure to clarify it and those who have mastered isolated concepts to begin to see how these concepts might fit together.

At times you may find yourself tempted to introduce an activity by giving examples or modeling how the activity might be accomplished. Avoid this. If you do this, you rob students of the chance to approach the activity in their own individual way. Instead of making assumptions about what students will or won't do, watch and listen. The excitement and challenge of the activity—as well as the chance to work cooperatively—may bring out abilities in students that will surprise you.

USING THE MANIPULATIVES

Each activity in this book was written with a specific manipulative in mind. The six manipulatives used are: Geoboards, Color Tiles, Snap Cubes, Cuisenaire Rods, Pattern Blocks, and Tangrams. All of these are pictured on the cover of this book. If you are missing a specific manipulative, you may still be able to use the activity by substituting a different manipulative. For example, most Snap Cube activities can be performed with other connecting cubes. In fact, if the activity involves using the cubes as counters, you may be able to substitute a whole variety of manipulatives.

The use of manipulatives provides a perfect opportunity for authentic assessment. Watching how students work with the individual pieces can give you a sense of how they approach a mathematical problem. Their thinking can be "seen" in the way they use and arrange the manipulatives. When a class breaks up into small working groups, you can circulate, listen, and raise questions, all the while focusing on how your students are thinking.

Work with manipulatives often elicits unexpected abilities from students whose performance in more symbolic, number-oriented tasks may be weak. On the other hand, some students with good memories for numerical relationships have difficulty with spatial reasoning and can more readily learn from free exploration with hands-on materials. For all students, manipulatives provide concrete ways to tackle mathematical challenges and bring meaning to abstract ideas.

Overview of the Lessons

INVESTIGATING AREA AND PERIMETER

OTHER AREA INVESTIGATIONS

INVESTIGATING VOLUME AND SURFACE AREA

Measurement, Grades 7-8

Investigating Area and Perimeter

The lessons in this cluster provide opportunities for students to explore areas of shapes that have the same perimeter, perimeters of shapes that have the same area, and the relationship between perimeter and the compactness of a shape. The activities reinforce understanding of area and perimeter and encourage creative exploration. They can be worked on in any order.

1. *Greta's Garden (Investigating areas of shapes that have the same perimeters)*

In Part 1 of this activity, students investigate the areas of possible rectangles that can be built using a given number of Color Tiles. Their investigation is couched in a problem involving the construction of a vegetable garden. Students compare their shapes and create a scale drawing of their garden model.

On Their Own Part 2 provides opportunities for students to explore other possible shapes that could be built using the same materials. Students explore how area is affected by changes made to a shape, and weigh the appeal of maximum area against the suitability of a particular shape for use as a garden.

2. Tiling Designs *(Investigating perimeters of shapes that have the same areas)*

In this lesson, students explore the range of possible perimeters that can be obtained by building shapes with a given area. In *On Their Own* Part 2, they use their discoveries to predict the range of possible perimeters that could be produced if the area of their shapes is increased by 2 square units. Students then test their predictions and look for ways to prove that their reasoning was valid.

The activities and discussion questions help students to clarify the concepts of area and perimeter. They also prompt students to consider the relationship between the two measurements. *Where's the Mathematics?,* (page 15) presents an idea for an extension investigation that leads to some interesting and unexpected results.

3. Sandboxes *(Exploring the relationship between area and perimeter)*

In this activity, students model rectangular sandboxes using a given set of Cuisenaire Rods. They devise ways for finding the areas of their models, and explore the range of possible areas and perimeters that can be obtained. As an extension, students can learn about a way to depict the results graphically, as presented in *Where's the Mathematics?* (page 20).

On Their Own Part 2 opens up the investigation, encouraging students to consider nonrectangular shapes for their sandboxes. Students design and build new models, and compare their areas, perimeters, and desirableness as shapes for sandboxes.

4. Perimeter Search *(Exploring perimeter as a function of compactness)*

In this activity, students use Pattern Blocks to explore perimeter. As they search to find all possible perimeters that can be made using a set of six blocks, students discover attributes of shapes that affect perimeter. Among their discoveries, is the concept that for shapes with the same area, compact shapes have smaller perimeters than elongated shapes.

On Their Own Part 2 provides students with an opportunity to apply their discoveries to a new set of blocks. They then have an opportunity to find shapes with the smallest and greatest perimeters in *For Their Portfolio.*

GRETA'S GARDEN

- Area
- Perimeter
- Scale drawing
- Spatial reasoning

Getting Ready

What You'll Need

Color Tiles, 60 per pair

Rulers

Color Tile grid paper, page 115

Activity Master, page 97

Overview

Students use Color Tiles to investigate ways to create a shape that has a given perimeter and the greatest possible area. In this activity, students have the opportunity to:

- recognize that shapes with the same perimeter may have different areas
- find ways to compare the area of different shapes
- draw a scale model
- reinforce their understanding of the concept of area

Other *Super Source* activities that explore these and related concepts are:

Tiling Designs, page 13

Sandboxes, page 17

Perimeter Search, page 22

The Activity

On Their Own (Part 1)

Greta wants to build a rectangular vegetable garden. She plans to surround the garden with square patio tiles. The box of tiles she bought contains 30 tiles, each measuring 2 feet on a side. What is the largest rectangular garden plot she can surround using these tiles?

- Working with a partner, use Color Tiles to build models of possible garden border tiles. Let each Color Tile represent 1 patio tile.
- Record your models and their dimensions on grid paper. Figure out and record the perimeter and area of each of your gardens. Remember to measure the perimeter and area of the *garden*, not the border.
- Select the model you would use if you were Greta. Create a scale drawing of the garden you choose.
- Be ready to explain why you selected this particular model.

Thinking and Sharing

Discuss the rectangular gardens that students modeled. Have pairs post drawings of the models they selected.

Use prompts like these to promote class discussion:

- How did you go about creating your models?
- How did you find the perimeters and areas of the rectangles you made?
- What made you choose the rectangles you chose?
- What did you notice about the areas of the different rectangles that could be surrounded by these same materials?

On Their Own (Part 2)

What if... ***the garden could be any shape? Could Greta build a bigger or better garden using the same box of patio tiles?***

- Consider other shapes that might be suitable for a garden.
- Build or sketch models of your ideas. You may use whatever materials you have available.
- Decide whether any of these gardens would be larger or more desirable than the rectangular garden you chose. Be ready to explain your reasoning.

Thinking and Sharing

Ask if any pairs think they have modeled a better garden. If so, have them post their drawings and explain why they chose their particular model.

Use prompts like these to promote class discussion:

- How did you go about modeling different-shaped gardens?
- How did you compare the areas of the different models you made?
- What did you notice about the areas of the different shapes you made?
- What made you choose the shape you chose?

Note that some of the students may decide that the best garden is not necessarily the one with the greatest area. For example, a student may choose a long narrow garden as being more desirable than a square one, as it allows the gardener easier access to the plants. For the same reason, another student may suggest building a tile path through the middle of the garden.

Write a brief letter to Greta describing the garden you think she should build and explaining why you think it is the best choice. Include any diagrams or instructions that might be helpful.

Teacher Talk

Where's the Mathematics?

Students may be surprised at the wide variety of shapes, both rectangular and nonrectangular, that have a given perimeter. In Part 1, students should find that the more compact their rectangle (that is, the closer the two dimensions), the greater its area. Three possible rectangular models are shown below.

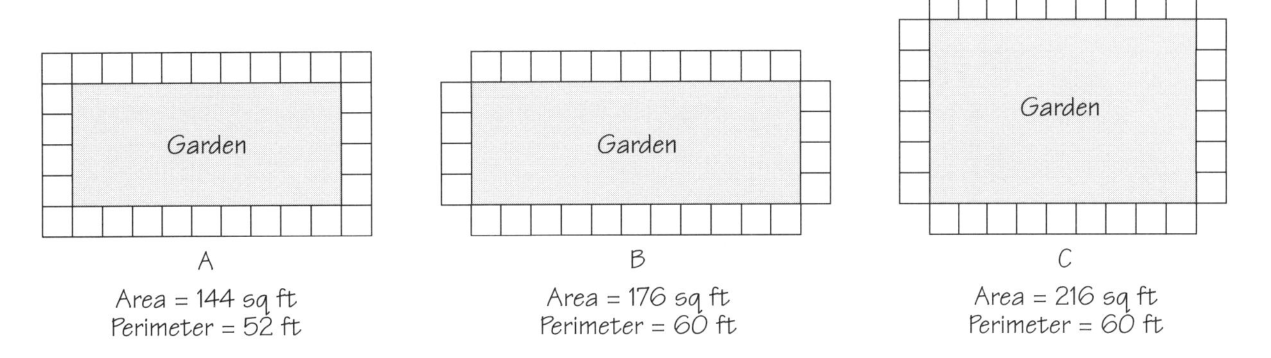

The models above show two different ways that tiles can be used to surround a rectangular garden. Some students may feel that the tiles should form a continuous path as they do in figure A; others may realize that they can increase the area of the garden by not placing tiles in the corners, as shown in figures B and C. Depending on how students arrange their patio tiles, the rectangular garden they find to have the greatest area will be either 16 ft by 14 ft (no corner tiles), or 14 ft by 12 ft (continuous tiles), as shown below.

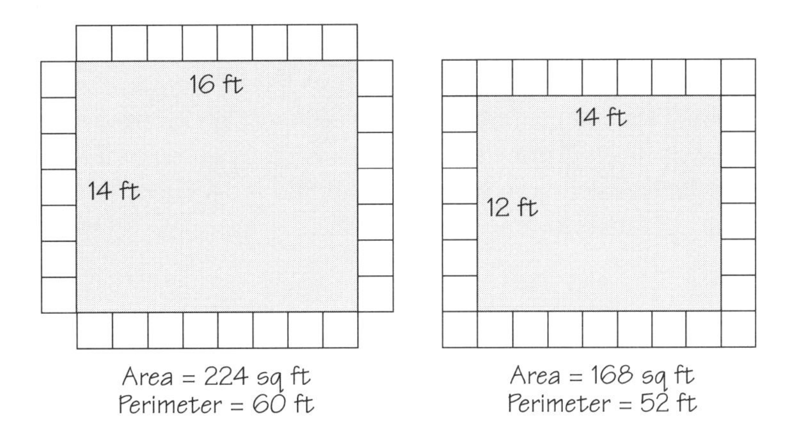

Students may recognize that a square is a type of a rectangle and may suggest that the garden with the greatest area would be square-shaped, measuring 15 ft on a side.

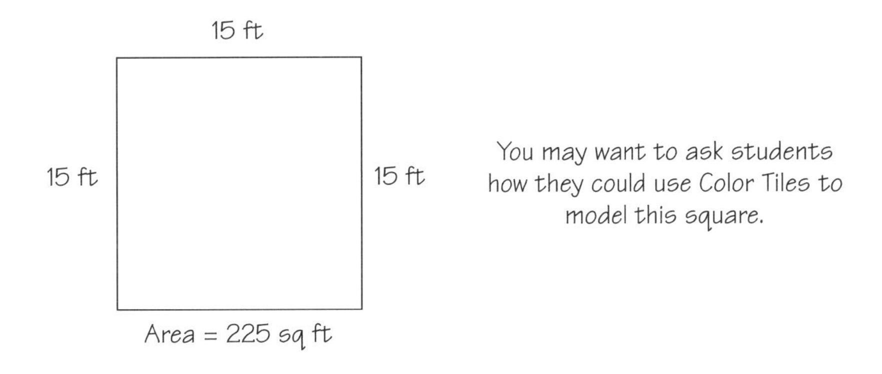

In Part 2, students may create a variety of different shapes for Greta's garden. Some may investigate polygons with many sides, such as octagons, decagons, or even 30-gons. Some examples are shown below.

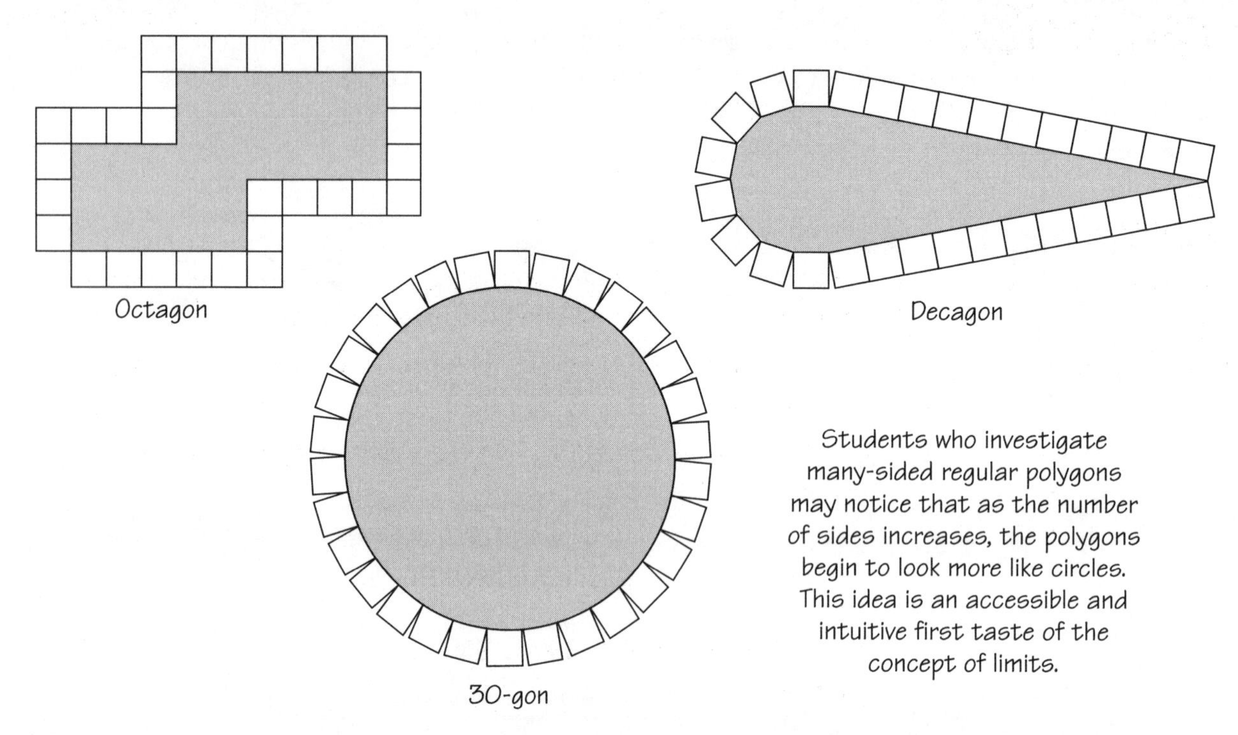

Some students may explore the possibility of building a circular garden. (Note: a circle with circumference 60 feet would have an area of approximately 286.5 sq ft, considerably larger than the area of any rectangular garden.) Others may stray from the original condition that the garden be surrounded by tiles and suggest enlarging the area by using an existing building, such as a house, for one side of the garden. To compare the areas of their gardens, some students may compare the numbers of Color Tiles that could be used to fill their shapes, approximating where necessary. Others may overlay their scale drawings on top of each other to compare them. Students need not find the exact areas of these shapes, but should instead focus on using spatial reasoning to compare their area to those of their other models.

TILING DESIGNS

- Area
- Perimeter
- Spatial reasoning

Getting Ready

What You'll Need

Color Tiles, about 30 per pair

Color Tile grid paper, page 115

Activity Master, page 98

Overview

Students use Color Tiles to investigate the perimeters of shapes that have the same area. In this activity, students have the opportunity to:

- recognize that shapes with the same area may have different perimeters
- explore the range of perimeters possible for shapes with a given area
- discover how the compactness of a shape effects its perimeter
- use deductive reasoning to prove a hypothesis

Other *Super Source* activities that explore these and related concepts are:

Greta's Garden, page 9

Sandboxes, page 17

Perimeter Search, page 22

The Activity

On Their Own (Part 1)

Marielle wants to tile the counter in her kitchen. She plans to use white tiles for most of the counter, and 26 colored tiles for a design near the center of the counter. In experimenting with different shapes, she made some interesting discoveries about the perimeters of the possible designs. What do you think she found?

- Working with a partner, make at least 10 tiling designs of different shapes, each using 26 Color Tiles. In each design, at least one complete side of each tile must touch at least one complete side of another tile.
- Record each of your designs on grid paper. Find and label the area and perimeter of each of your shapes.
- Be ready to discuss your findings about the perimeters of your designs.

Thinking and Sharing

Invite students to share their results and to talk about their discoveries. Ask the pair who made the shape with the smallest perimeter to post their design on the left side of the chalkboard. Have students post other shapes across the board, ranging from the smallest to the greatest perimeter.

Use prompts like these to promote class discussion:

- What did you notice about the areas of the shapes you made?
- What did you notice about the perimeters of the shapes you made?
- How can the shapes all have the same area, and yet most of them have different perimeters?
- How would you compare the shapes that have the smallest perimeters to those that have the greatest perimeters?
- Do you think the perimeters of the posted shapes are all the perimeters that are possible to make using the 26 tiles? If not which ones are missing? How do you know?

On Their Own (Part 2)

What if... *the design is to be made from 28 square tiles? What could you predict about the range of possible perimeters of the different shapes that could be made?*

- With your partner, discuss and record a prediction about the possible perimeters of shapes made with 28 tiles.
- Use Color Tiles to test your prediction. Modify your hypothesis if necessary.
- Be ready to prove that your hypothesis is correct.

Thinking and Sharing

Invite students to share their hypotheses and present their proofs.

Use prompts like these to promote class discussion:

- How did you go about making your original prediction?
- Did your work with the Color Tiles cause you to modify your hypothesis? If so, how?
- Were you surprised by your findings? Explain.
- How were you able to prove that your hypothesis was correct?
- How might you generalize your findings?

Write a brief letter to Marielle describing the discoveries you made about the perimeters of shapes containing 26 tiles. Be sure to explain why only certain perimeters are possible.

Teacher Talk

Where's the Mathematics?

In working through problems of the type presented in this lesson, students have an opportunity to clarify the concepts of area and perimeter. They can also see that not all shapes with the same area have the same perimeter. The areas of shapes made from the 26 tiles are 26 square units, whereas the perimeters range from 22 to 54 units (assuming tiles are joined as instructed).

Students may observe that the shapes with greater perimeters are often more spread out than those with smaller perimeters, which tend to be more compact. For example, the elongated shape in Figure A has a perimeter of 34 units, while the more compact shape in Figure B has a perimeter of 26 units.

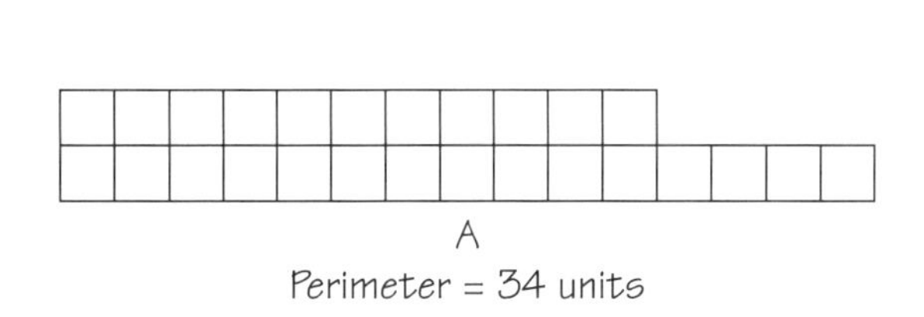

A
Perimeter = 34 units

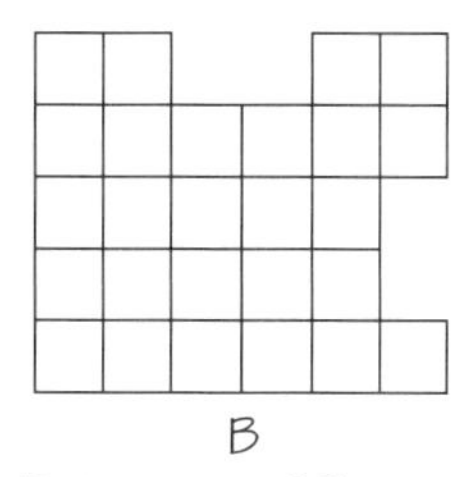

B
Perimeter = 26 units

Students may reason that in shapes that are compact, tiles share more sides with each other than in those that are more elongated. Thus, in the compact shapes, there are fewer tiles exposed to the outside of the shape contributing to the perimeter. In this example, each tile marked with an "x" is completely surrounded by other tiles, and therefore contributes nothing to the perimeter.

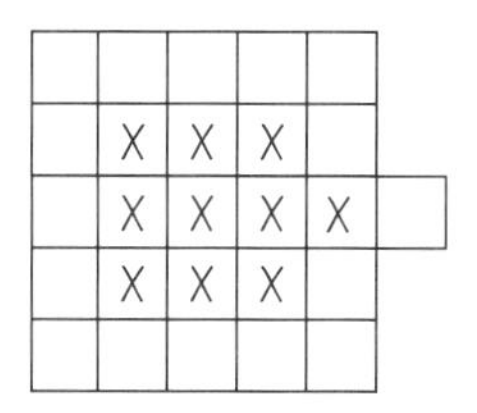

In the elongated shape below, there are no such "interior" tiles; every tile is exposed to the outside of the shape and contributes at least two units to the perimeter.

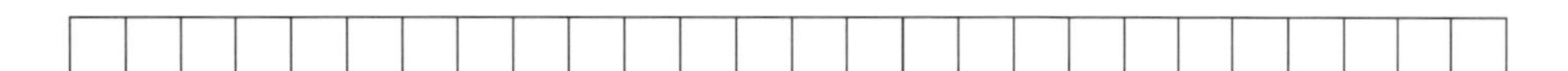

Students may observe that the perimeters are all even numbers of units. In fact, if the class results show a pattern of even numbers and one or more is missing, students may be motivated to try to create shapes with the missing perimeters. Students may also want to investigate to see if the smallest and greatest perimeters identified by the class are actually the smallest and greatest perimeters possible. Their work should verify that the possible perimeters are all of the even numbers between and including 22 and 54.

Students may have different ways of explaining why the perimeters are all even. One interesting explanation involves comparing the numbers of units in the top and bottom edges, and those in the left and right edges of each shape. For example, in the shape shown here, the edges marked with a "t" can be considered "top edges" relative to those marked with a "b," which can be considered "bottom edges." Similarly, those edges marked with an "l" can be considered "left edges" relative to those marked with an "r," which can be considered "right edges." Notice that the number of units in the top edges is equal to the number of units in the bottom edges, and the number of units in the left edges is equal to the number of units in the right edges.

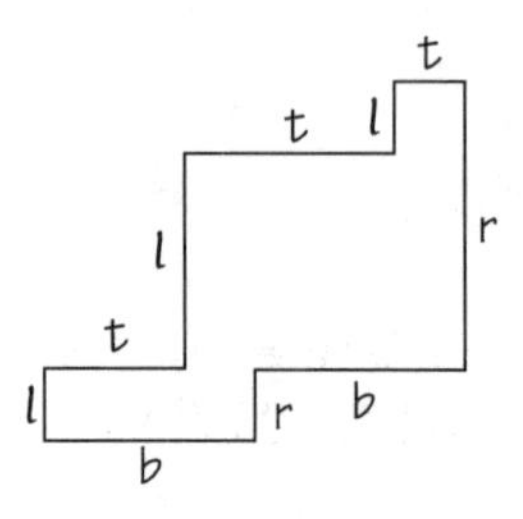

Thus, the number of units in the horizontal edges will always be a multiple of two as will the number of units in the vertical edges. The total number of units (the perimeter) will therefore always be the sum of two even numbers, which is even.

In Part 2, students may hypothesize that the smallest and greatest perimeters will increase by 2 or by 4 once two tiles are added. Students may try to test their predictions by building the most elongated and most compacted shapes possible with 28 tiles. Students may be surprised to find that the possible perimeters range from 22 to 58 units. The most surprising aspect of this discovery may be that the smallest possible perimeter (created by the more compact shapes) is the same as the smallest possible perimeter that could be created with 2 fewer tiles. Students may want to explore why this is so by experimenting to see how many tiles are needed to produce a situation in which the smallest possible perimeter is greater than (or less than) 22 units. Such an investigation will provide thought-provoking results and may lead to a greater understanding of the relationship between area and perimeter.

SANDBOXES

- Area
- Perimeter
- Spatial reasoning

Getting Ready

What You'll Need

Cuisenaire Rods, 1 set per pair

1-centimeter grid paper, page 116

Activity Master, page 99

Overview

Students investigate area and perimeter by modeling different-shaped sandboxes using Cuisenaire Rods. In this activity, students have the opportunity to:

- devise strategies for finding area and perimeter
- investigate how area can change while perimeter remains constant
- explore the range of areas possible for shapes with a given perimeter

Other *Super Source* activities that explore these and related concepts are:

Greta's Garden, page 9

Tiling Designs, page 13

Perimeter Search, page 22

The Activity

On Their Own (Part 1)

Students from the Longview Middle School want to build a sandbox for the local playground. Using the wood they have available, how many different-sized rectangular sandboxes can they build?

- Use the following set of Cuisenaire Rods to represent the wood they have available: 2 green, 2 purple, 2 yellow, 2 dark green, and 2 black.
- Working with a partner, build models of possible rectangular sandboxes. Each sandbox must use all 10 rods for the surrounding wall. Where rods meet, they must touch along one square centimeter, not just corner to corner. Try to find sandboxes in every possible size.

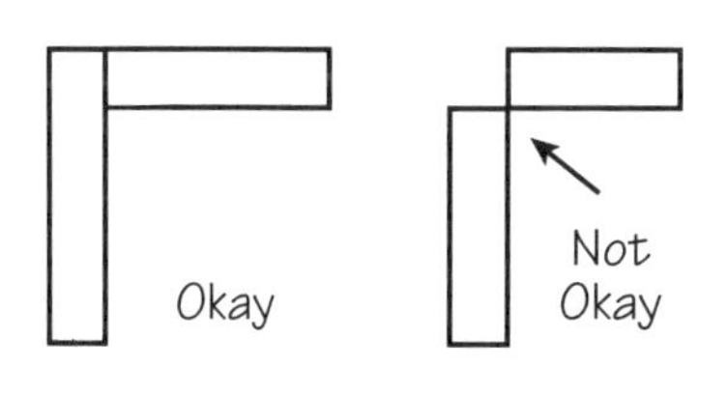

- Record your sandboxes on grid paper. Using the length of one white rod to represent 1 foot, find and record the length, width, area, and perimeter of the play area *inside* each of your sandboxes.
- Select the model you think the students should use for the sandbox. Be ready to explain why you selected this particular model.

Thinking and Sharing

Discuss the students' sandbox models. Compile a class chart listing the areas from smallest to greatest. Next to each area, list the corresponding dimensions and the perimeter.

Use prompts like these to promote class discussion:

- How did you go about creating your models?
- How did you find the perimeters and areas of the sandboxes you made?
- What did you notice about the areas of the different rectangles that could be surrounded by these same materials?
- Do you think you found all the possible rectangular sandbox sizes that can be made using the given set of rods? How do you know?
- What criteria did you use for selecting one model as the "best" one for students to use?
- What patterns do you notice in the class chart?

On Their Own (Part 2)

What if... ***the sandbox could be any shape? Could the same set of materials be used to build a sandbox that might be more desirable than a rectangular one?***

- Working with your partner, consider other shapes that may be suitable for a sandbox.
- Using the same set of rods, build and sketch models of your ideas. (Note: As in Part 1, rods much touch along one square centimeter and you must use all 10 rods for the surrounding wall.)
- Decide whether any of these sandboxes would be more desirable than the rectangular sandbox you chose in Part 1. Be ready to explain your reasoning.

Thinking and Sharing

Ask if any pairs think they have modeled a better sandbox. If so, have them post their drawings and explain why they chose their particular model.

Use prompts like these to promote class discussion:

- How did you go about modeling different-shaped sandboxes?
- How did you compare the areas of the different models you made?
- What made you choose the shape that you chose?

Write a summary describing what would happen if there were no restrictions on the shape of your sandbox and your rods did not have to touch along one square centimeter? How would these changes effect the area and perimeter of your sandboxes? Decide which shapes(s) would be preferable for a sandbox and explain why.

Teacher Talk

Where's the Mathematics?

As they work with the given set of rods, students discover that they can build many different rectangles that have the same perimeter, but that differ in area.

Students' previous exposure to the concepts of perimeter and area may vary greatly. Consequently, students may go about this activity in different ways. Some may rely on counting and adding the lengths of the Cuisenaire Rods to find the perimeter and counting the squares on the grid paper to find the area. Others may use formulas to find these measurements. For some students, this activity will provide an experience that will explain why the formula for perimeter is $P = 2l + 2w$ [or $P = 2(l + w)$] and the formula for area is $A = lw$.

This activity should help students overcome some common misconceptions. Initially, students may be surprised to find that all the rectangles have the same perimeter. Students frequently assume that long, skinny rectangles have greater perimeters than those that are more squared off. Students may also assume that there is a direct relation between area and perimeter and that as area increases, so must the perimeter. Their work should convince them that this is not the case.

By compiling the data in a class chart, students can analyze their results from an algebraic point of view. When the data are arranged according to the area measurements (as shown), students can notice that as one of the dimensions increases, the other decreases, until they are nearly the same. As this happens, the shape of the rectangle approaches that of a square. Students can also conclude that only 11 unique rectangles are possible. The next entry in the chart would be a rectangle with dimensions 12 x 11, the same as the last rectangle listed. This indicates that successive rectangles would be duplications of those already listed.

Dimensions	Perimeter (feet)	Area (square feet)
1 x 22	46	22
2 x 21	46	42
3 x 20	46	60
4 x 19	46	76
5 x 18	46	90
6 x 17	46	102
7 x 16	46	112
8 x 15	46	120
9 x 14	46	126
10 x 13	46	130
11 x 12	46	132

Students should observe that the rectangle with the greatest area is the one that is most nearly square, and that longer, narrower rectangles have smaller areas. The chart also reveals that as the rectangles grow more square-like, the areas increase by consecutively smaller even numbers. For example, the area of the second rectangle is 20 square feet more than the first; the area of the third is 18 square feet more than the second; the area of the fourth is 16 square feet more than the third; and so on.

You may want to extend the algebraic analysis one step further and have students construct a graph of the area as the length increases from 1 foot to 22 feet. The graph provides a nice example of a parabola and gives students experience in graphing a nonlinear relationship [$A = l(23 - l)$].

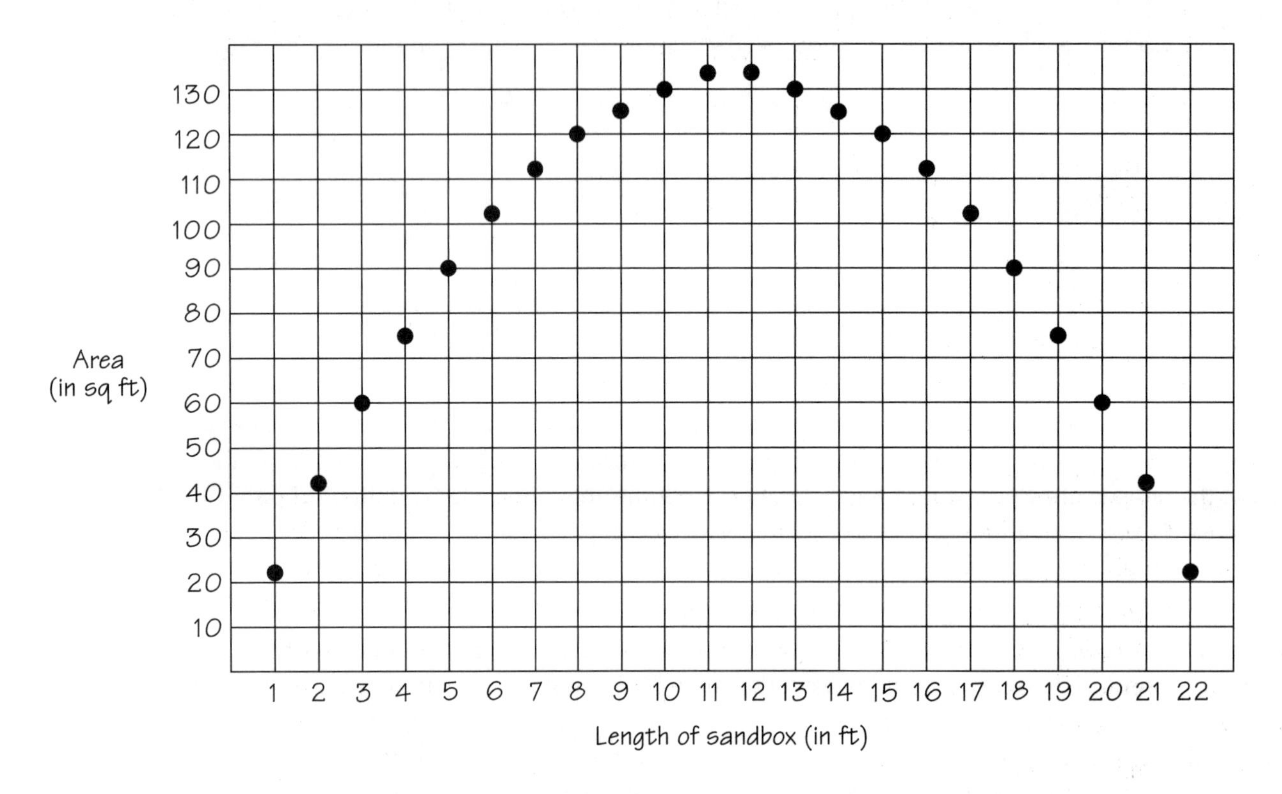

In Part 2, if students continue to follow the rule that the rods must touch along 1 square centimeter, they will find that the perimeter of any sandbox they create will be 46 feet and that the areas will still be in the range of 22 to 132 square feet. There are a multitude of shapes possible. Some examples are shown.

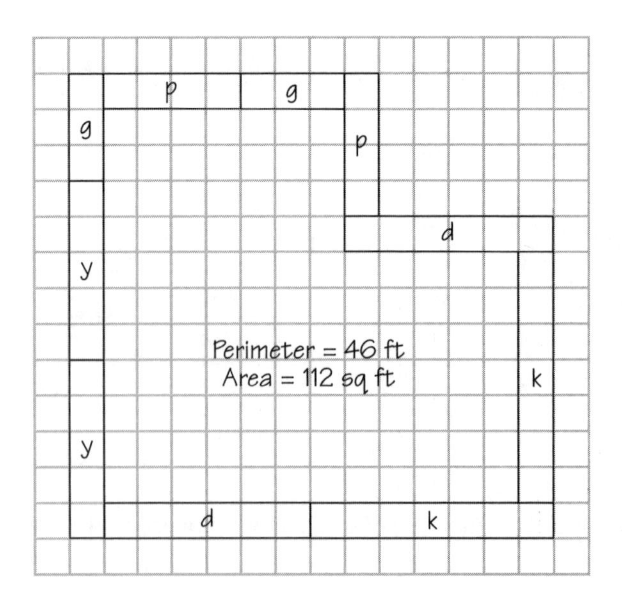

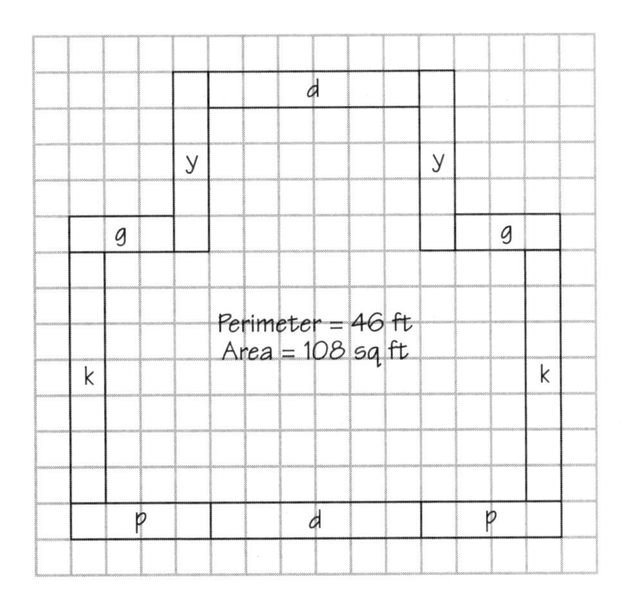

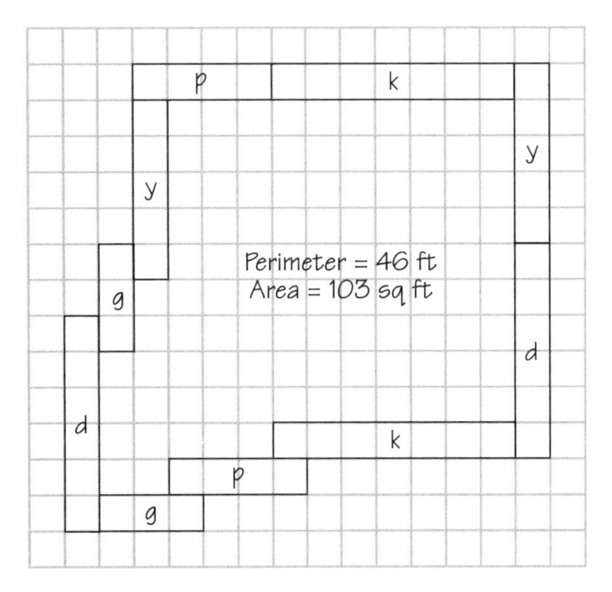

Although the shapes above do not provide additional perimeter or area, students may suggest that one of these designs might be more aesthetically pleasing than a rectangular sandbox. They may also decide that an irregular shape allows for bordered play areas of different sizes.

In *For Their Portfolio*, some students may find that they can build a sandbox with an area greater than 132 square feet when they disregard the restriction about how the rods must be joined. For example, the decagon model below has a perimeter of 50 feet and an area exceeding 175 square feet.

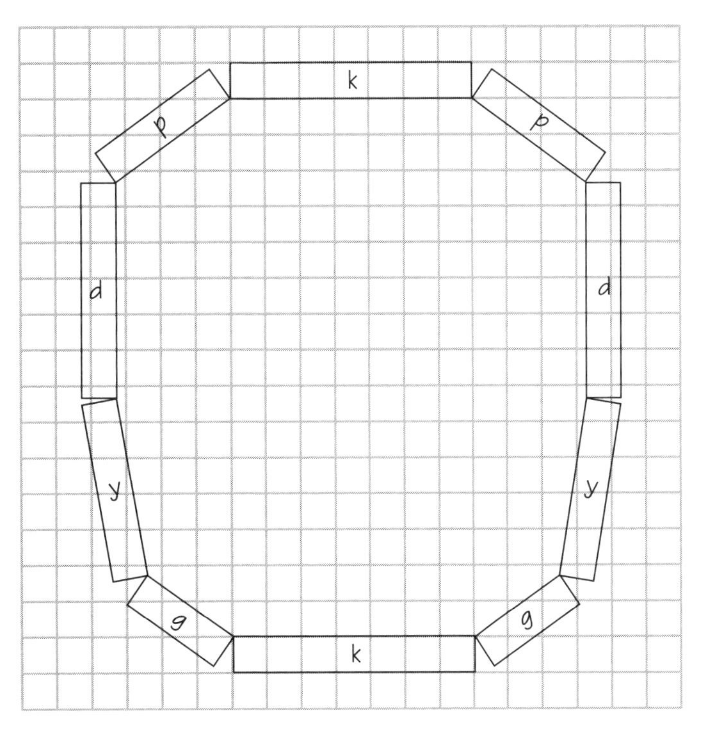

Some students may argue that a sandbox built from wood joined corner to corner would not be desirable because the structure would be unstable. They may suggest overlapping the ends to some degree of support. Students should recognize how such a change effects both area and perimeter.

Students may have to rely on counting unit squares to find the areas of some of the irregular shapes they make. Some students may divide the interiors of their shapes into small rectangles whose areas can be determined using the formula $A = lw$, and then estimate the number of squares in the remaining parts of the interior.

As it is not necessary for students to find the exact areas of their sandboxes, some students may use combinations of comparison, estimation, and spatial reasoning skills to make judgments about their different models.

PERIMETER SEARCH

- Perimeter
- Spatial reasoning
- Nonstandard units of measure

Getting Ready

What You'll Need

Pattern Blocks, at least 6 of each shape per pair

Pattern Block triangle paper, page 117

Crayons or markers

Activity Master, page 100

Overview

Students investigate the perimeters of shapes that can be made using different combinations of Pattern Blocks. In this activity, students have the opportunity to:

- reinforce their understanding of the concept of perimeter
- devise strategies for creating shapes with different perimeters
- explore the relationship between the compactness of a shape and its perimeter

Other *Super Source* activities that explore these and related concepts are:

Greta's Garden, page 9

Tiling Designs, page 13

Sandboxes, page 17

The Activity

On Their Own (Part 1)

During National Mathematics Week, Mr. Frangione invited students to present challenging problems to the class. Debbie posed the following question:* What are all the possible perimeters of shapes that can be made using six Pattern Blocks? *Can you solve Debbie's problem?

- Working with your partner, find all of the possible perimeters of shapes that can be made using six Pattern Blocks. You may use as many different combinations of blocks as you like.
- Use the length of the side of a green triangle as the unit of measure. Be sure to fit your blocks together so that each block shares at least one unit of length and at least one vertex with another block.
- Copy your shapes onto triangle paper and record the perimeter of each shape.
- Be ready to discuss the strategies you used to make shapes with new perimeters.

Thinking and Sharing

Ask students to tell what perimeters they found. List the perimeters on the chalkboard in numerical order. Invite pairs to share and compare the shapes they made and the strategies they used for creating shapes with new perimeters.

Use prompts like these to promote class discussion:

- How did you go about creating shapes with different perimeters?
- How did you search for the shape with the smallest perimeter? Do you think it is possible to make other shapes with this same perimeter? Why or why not?
- How did you search for the shape with the greatest perimeter? Do you think it is possible to make other shapes with this same perimeter? Why or why not?
- What generalizations can you make about the characteristics of shapes having large (small) perimeters?
- How do you know that you have found all possible perimeters?

On Their Own (Part 2)

What if... ***your shapes must be made using one of each of the six different Pattern Blocks? What perimeters would be possible?***

- Investigate the perimeters of shapes that can be made using one of each of the six different Pattern Blocks.
- Record your shapes and their perimeters.
- Be ready to discuss and explain your findings.

Thinking and Sharing

Invite students to tell about the shapes they made and their perimeters.

Use prompts like these to promote class discussion:

- What did you discover about the perimeters of shapes that could be made with the six different blocks?
- How did you go about searching for shapes with different perimeters?
- Why do you think the possible perimeters are limited to the ones you found?

Describe how you would go about finding the shapes with the smallest and greatest perimeters if you were given a set of ten Pattern Blocks.

Teacher Talk

Where's the Mathematics?

In their search for all possible perimeters, students may realize that they can create the smallest possible perimeter using the smallest blocks (the green triangles) and the greatest possible perimeter using the largest blocks (the yellow hexagons). It is through their work with these blocks that they may discover that the blocks need to be arranged in certain ways to make shapes with the absolute smallest (and greatest) perimeters. For example, although several different shapes can be made using six green triangles, only the regular hexagon pictured below has the smallest possible perimeter, 6 units.

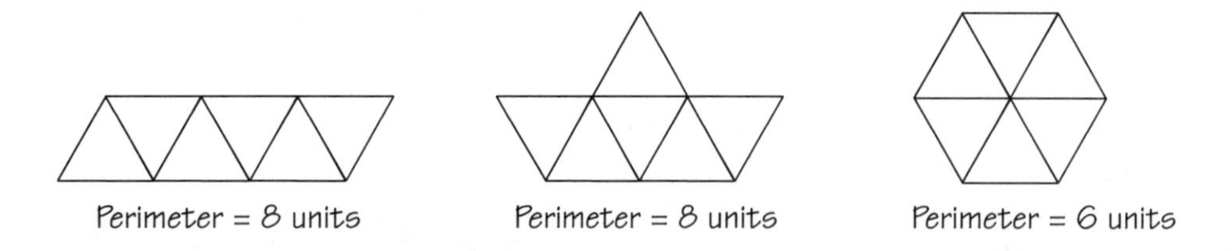

Likewise, six hexagons can be arranged in a number of ways, but only some of them yield the greatest possible perimeter, 26 units.

Perimeter = 18 units

Perimeter = 20 units

Perimeter = 22 units

Perimeter = 26 units

Perimeter = 26 units

Some students may use trial and error to build shapes that have different perimeters, while others may use a more systematic approach. For example, students may begin with a particular shape, calculate its perimeter, and then exchange one or more blocks in the shape for blocks that will increase (or decrease) the perimeter. An example of one use of this strategy is shown on the next page.

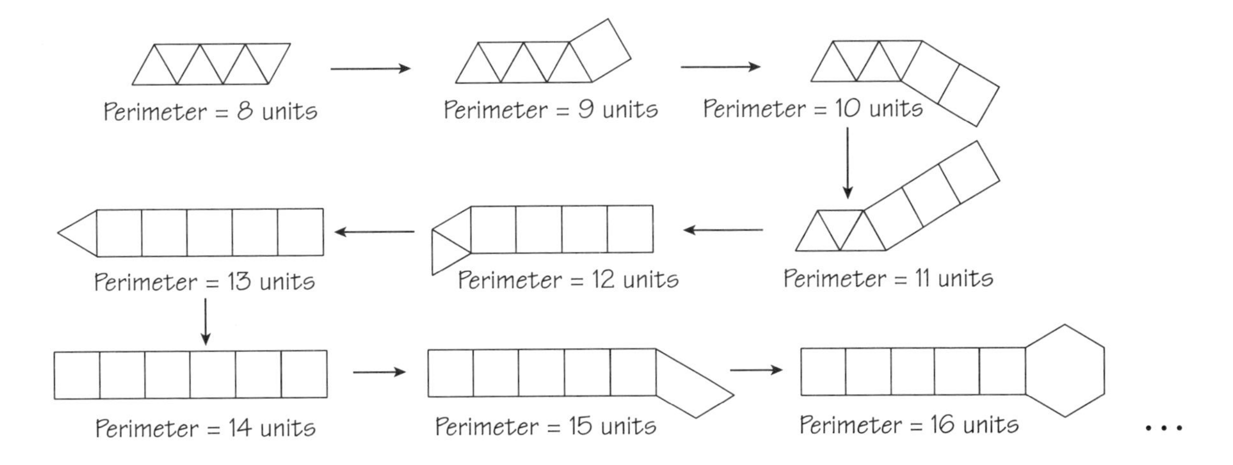

As students experiment with different ways of arranging the blocks, they may discover that shapes that are more compact usually have smaller perimeters than those that are more elongated. Students may use this observation to build and alter their shapes to produce new shapes with greater or smaller perimeters. Although there is only one way to make a shape that has the smallest perimeter, students should find that there are multiple ways of arranging the blocks to build shapes that have other perimeters. Students may notice that perimeters that lie towards the middle of the range of possible perimeters can be obtained by arranging the shapes in the widest variety of different ways.

In Part 2 of the activity, students may be surprised to find that only three perimeters are possible to obtain using six different Pattern Blocks: 12 units, 14 units, and 16 units. Some examples are shown here.

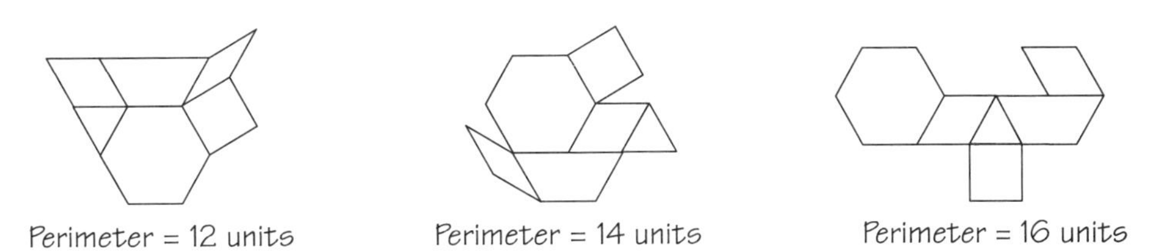

Perimeter = 12 units   Perimeter = 14 units   Perimeter = 16 units

Students may use what they learned from Part 1 of the activity, building compact shapes and elongated shapes, to investigate the smallest and greatest possible perimeters. As they rearrange the blocks to try to create shapes with different perimeters, students can see how the perimeter is affected by sides that become exposed or concealed by the change in arrangement. For example, when the orange square in Figure A is relocated as shown in Figure B, it will have one additional side exposed to the perimeter. Furthermore, the side of the trapezoid and the side of the tan rhombus that were previously concealed will also be exposed. With the concealment of one additional side of the blue rhombus, the net change is an increase of two units of perimeter.

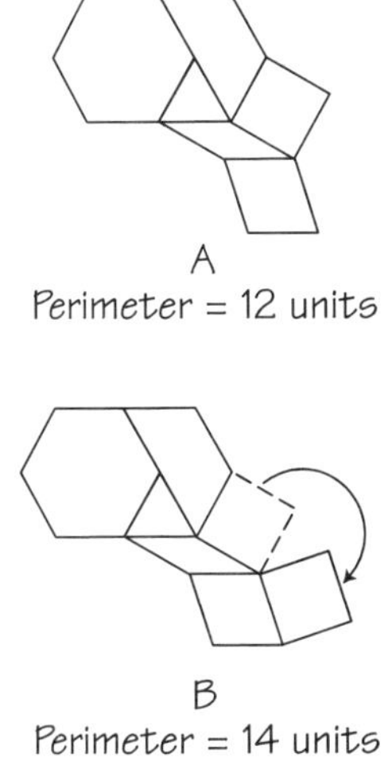

Students may notice that whereas in the first investigation it was only possible to find one shape with the smallest perimeter and one with the greatest, in Part 2 this is not the case. This observation may motivate students to think about the different restrictions that were imposed in the two problems and how they affected the results.

Other Area Investigations

The lessons in this cluster give students the opportunity to investigate properties of similar figures with respect to side lengths, perimeters, and areas, changes in area or volume resulting from changes in the dimensions of the shape or solid, Pick's theorem for computing areas, and areas expressed in terms of other areas.

1. Stadium Flip Cards *(Investigating relationships between similar figures)*

In this activity, students explore the relationships that exist among similar shapes. Their investigation involves a problem in which they use Color Tiles to represent models of stadium flip card letters. Using ratios and proportions, students determine the area and perimeter of the actual stadium card letters based on their models.

In *On Their Own* Part 2, students search for relationships between various scale factors and changes in areas and perimeters of different stadium letters. The lesson serves as an introduction to the ratios between side lengths, perimeters, and areas of similar figures.

2. Bon Voyage *(Investigating changes in area)*

In *On Their Own* Part 1, students use Color Tiles to investigate how changes in the length and/or width of a rectangle affect its area. Students organize their data to see relationships between changes in dimensions and corresponding changes in area.

In *On Their Own* Part 2, students use Snap Cubes to extend this concept to changes in volumes of solids caused by changes to one, two, or three dimensions.

3. Wholes and Holes *(Exploring areas of polygons using Pick's theorem)*

Before introducing this activity, teachers may want to review the location of border pegs and interior pegs on the Geoboard and Pick's theorem, which is used to find the area of a Geoboard triangle.

In this activity, students build Geoboard polygons and determine their areas using a variety of mathematical approaches. Some students may use a partitioning technique, while others may choose to find the areas of the regions surrounding their polygon and subtract their total from the area of the enclosing square.

In Part 2, students find the area of a "donut," the region enclosed between two polygons. Interesting discussion may result when students test Pick's theorem on the entire "donut."

4. Puzzle in a Puzzle *(Exploring areas of similar figures)*

Students are challenged to arrange different numbers of Tangram pieces to generate shapes similar to a given figure and express their side lengths and areas in terms of the original figure's side length and area.

In *On Their Own* Part 2, students are asked to build a hexagonal figure with Pattern Blocks and determine the fractional part or percent that each shape represents in relation to the whole.

STADIUM FLIP CARDS

- Scale drawing
- Ratios and proportions
- Perimeter
- Area

Getting Ready

What You'll Need

Color Tiles, 150 Color Tiles per pair

Color Tile grid paper, page 115

Activity Master, page 101

Overview

Starting with models of four letters made from Color Tiles and the total area of similar larger letters, students determine the scale factor used to generate these letters and their perimeters. In this activity, students have the opportunity to:

- explore ratios used in scale drawings
- use proportions to solve problems
- formulate the relationships between ratio of sides, ratio of perimeters, and ratio of areas
- investigate the concept of similar figures and scale factors

Other *Super Source* activities that explore these and related concepts are:

Bon Voyage, page 32

Wholes and Holes, page 38

Puzzle in a Puzzle, page 43

The Activity

On Their Own (Part 1)

Ruben's school is hosting a Math Olympiad. Ruben wants to make stadium flip cards that can be used by the students in the bleachers to cheer on the teams. Each student will hold 1 flip card. Can you help Ruben use a Color Tile model of the word "MATH" to determine the number of cards needed to form each flip card letter, the scale factor used, and the perimeter of each flip card letter?

- Work with a partner. Use Color Tiles to make a model of the word "MATH." Model letters must be 5 tiles wide by 5 tiles tall.

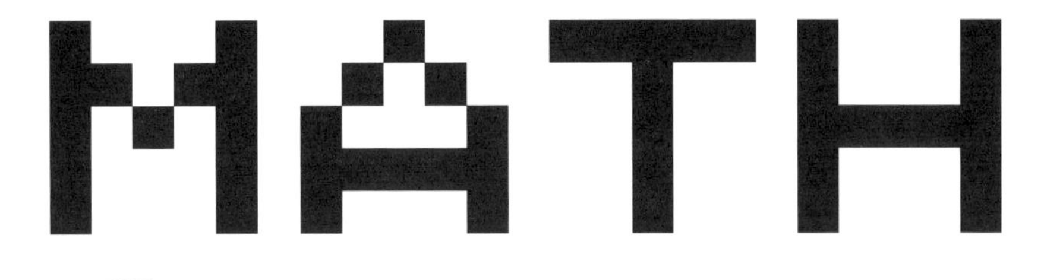

- Find the number of square units in the area of each letter, the total number of Color Tiles used, and the perimeter of each letter. Organize and record your results on a chart.
- Consider this tile arrangement to be a scale version of the students' "MATH" flip cards. If 423 students will hold cards, determine how many students are needed to hold flip cards for each of the four letters. Record these results on your chart.
- Build one or more of these new flip card letters using Color Tiles. Make sure that it is similar in design to its model. Find the perimeter(s). Add these results to your chart.
- What patterns or relationships did you find? Be ready to discuss your findings.

Thinking and Sharing

Compile students' results into a chart with the following headings: *Model Letter, Model Area, Model Perimeter, Area of Flip Card Letter.* When students refer to the different ratios during the discussion, the concept of "scale factor" may be introduced.

Use prompts like these to promote class discussion:

- What is the area and perimeter of each scale model letter?
- How did you determine the area of each stadium flip card letter knowing the total number of students?
- How does the area of each model letter compare to the area of the corresponding flip card letter?
- Is the ratio for each pair of letters (the model letter and the flip card version) consistent with the ratio for all other pairs of matching letters? Why or why not?
- What patterns or relationships did you discover?
- Did you come up with a similar relationship for the ratios of perimeters? Explain.

On Their Own (Part 2)

What if... ***Ruben wants to create a series of flip cards that spell out "EASY AS PI"? He must first build a Color Tile model of each letter and then, based on the scale factors he wants to use, determine the area and perimeter for each of these flip card letters. (The Color Tile model letters must be 5 tiles wide by 5 tiles tall). Can you help him with this task?***

- Using Color Tiles, build models of the letters used in the phrase "EASY AS PI."
- Find the area and perimeter of each letter. Make a chart of your findings.
- Using Color Tiles, create similar flip card letter models with a scale factor of 2.

- Find the areas and perimeters of these flip card letters and add this data to the chart under a column labeled *Scale Factor: 2.*
- Using the original Color Tile model, repeat the process to find the scale factors of 3, 4, 5, and 6, where possible. Record your findings on the chart and write about any relationships you notice in the data.
- Based on what you discovered above, consider a scale factor of *n* and *predict* the area and perimeter of the flip cards needed for the letters in "EASY AS PI."
- Be ready to discuss your findings.

Thinking and Sharing

Invite students to illustrate their Color Tile letter arrangements on the board. Have students develop a chart similar to that used in the first activity to record the data.

Use prompts like these to promote class discussion:

- How did you go about designing the letters used in the phrase "EASY AS PI"?
- What was the area and perimeter for each letter?
- How did you go about building the new letters using a scale factor of 2?
- What was the new perimeter and area for each letter?
- How can knowing a scale factor help in determining the perimeter and area of each newly formed letter?
- What were the perimeters and areas for letters whose scale factors were 3, 4, 5, and 6?
- What affect on the area and perimeter of the original letters does a scale factor of 3 have? a scale factor of 4? a scale factor of 5? a scale factor of 6?
- How did you represent the area and perimeter of a new letter formed using a scale factor of *n*?

Write a note to Ruben describing the change in area for a flip card letter (or any geometric shape) if the scale factor is between 0 and 1, equal to 1, or greater than 1. Include any diagrams that would be helpful.

Teacher Talk

Where's the Mathematics?

As students begin to consider the tasks of finding the area and perimeter of each letter based on the total number of flip cards used, they will need to use ratios and proportions. By counting the number of Color Tiles used to form each letter in the word "MATH," and by comparing that number to the total number of tiles used, students will be able to determine the number of flip cards used to form the corresponding letters by the 423 students in the stadium. The results are shown below:

"M"	"A"	"T"	"H"
Area = 13 Color Tiles	Area = 12 Color Tiles	Area = 9 Color Tiles	Area = 13 Color Tiles
Perimeter = 32 units	Perimeter = 32 units	Perimeter = 20 units	Perimeter = 28 units

Total number of Color Tiles used is 47.

Solving the following proportions for the value of the unknown, students are able to calculate the number of stadium flip cards used to form each letter.

"M"	"A"	"T"	"H"
$\frac{13}{47} = \frac{x}{423}$	$\frac{12}{47} = \frac{x}{423}$	$\frac{9}{47} = \frac{x}{423}$	$\frac{13}{47} = \frac{x}{423}$
$47x = (13)(423)$	$47x = (12)(423)$	$47x = (9)(423)$	$47x = (13)(423)$
$x = 117$ flip cards	$x = 108$ flip cards	$x = 81$ flip cards	$x = 117$ flip cards
Ratio of areas 13 : 117 = 1 : 9	Ratio of areas 12 : 108 = 1 : 9	Ratio of areas 9 : 81 = 1 : 9	Ratio of areas 13 : 117 = 1 : 9

At this point, students may attempt to calculate the perimeter of each flip card letter. They may believe that since the ratio of the areas is 1:9, the ratio of the perimeters might be the same. However, when students notice that the number of Color Tiles used to form "M" equals the number of Color Tiles used for "H," they should also note that their perimeters are different, thus, making it inappropriate to apply the ratio of 1:9 to find the perimeters of the flip card letters.

When students have completed building one or more of the flip card letters and have determined the perimeter(s), they will discover that the ratio of the perimeter of the model letter to that of the flip card letter always simplifies into the ratio 1:3.

Ratio of perimeters: "M" 32:96 "A" 32:96 "T" 20:60 "H" 28:84

When the ratio of the perimeters is compared to the ratio of the areas, students will notice that the first ratio is the square of the second. That is, $(1:3)^2 = 1:9$. One way to justify this relationship, is to have students consider the number of dimensions represented by each measurement. The perimeter is a one-dimensional measurement-length, while the area is a two-dimensional measurement-length and -width. As each letter undergoes a size change, students need to remember that both the perimeter and area undergo different, but related changes. It may be easier for the student to understand the concept using one square and the corresponding alterations in its perimeter and area if the side length is multiplied by a scale factor of 3.

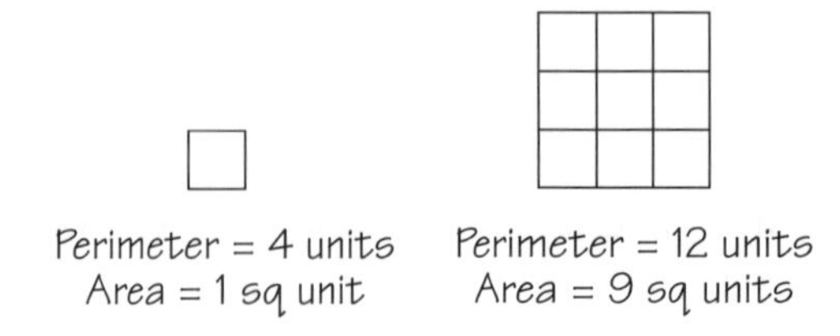

In the second activity, students are asked to build Color Tile models for the letters used in the phrase "EASY AS PI." The letter designs shown below might be possible arrangements for the tiles.

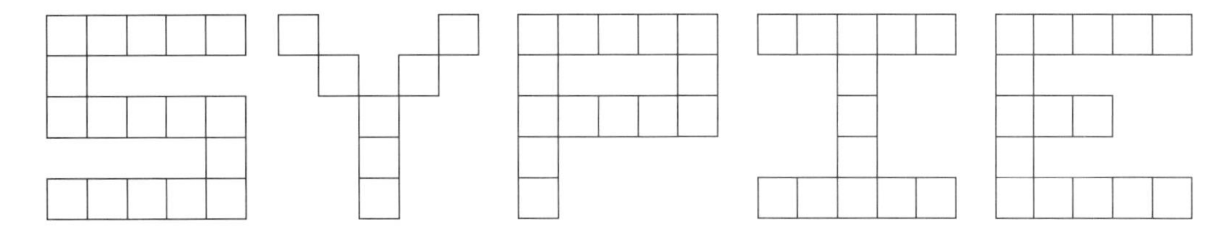

The chart below lists the area and perimeter for each Color Tile model letter shown above. When students build flip card letters similar to their models using a scale factor of 2, they will find that both the length and the width are doubled. Thus, the perimeter of the new flip card letter formed will be twice the original model's perimeter but the area of the new flip card letter will be 2 x 2 = 2^2, or 4 times larger than the area of the original tile model. When the scale factor is changed to 3, the flip card's perimeter will be tripled and its area will be multiplied by 3 x 3 = 3^2 or 9. When the scale factor is increased to 4, the flip card's perimeter will be quadrupled and its area will be multiplied by 4 x 4 = 4^2 or 16, and so on.

Letter	Color Tile Model	Scale Factor 2	Scale Factor 3	Scale Factor 4	Scale Factor 5	Scale Factor 6	Scale Factor *n*
E	Perimeter = 32	P = 64	P = 96	P = 128	P = 160	P = 192	P = 32*n*
	Area = 15	A = 60	A = 135	A = 240	A = 375	A = 540	A = $15n^2$
A	Perimeter = 32	P = 64	P = 96	P = 128	P = 160	P = 192	P = 32*n*
	Area = 12	A = 48	A = 108	A = 192	A = 300	A = 432	A = $12n^2$
S	Perimeter = 36	P = 72	P = 108	P = 144	P = 180	P = 216	P = 36*n*
	Area = 17	A = 68	A = 153	A = 272	A = 425	A = 612	A = $17n^2$
Y	Perimeter = 24	P = 48	P = 72	P = 96	P = 120	P = 144	P = 24*n*
	Area = 7	A = 28	A = 63	A = 112	A = 175	A = 252	A = $7n^2$
P	Perimeter = 28	P = 56	P = 84	P = 112	P = 140	P = 168	P = 28*n*
	Area = 14	A = 56	A = 126	A = 224	A = 350	A = 504	A = $14n^2$
I	Perimeter = 28	P = 56	P = 84	P = 112	P = 140	P = 168	P = 28*n*
	Area = 13	A = 52	A = 117	A = 208	A = 325	A = 468	A = $13n^2$

As successively larger flip card letters are considered and it becomes increasingly more difficult to build them using Color Tiles, it may be necessary for students to rely on the data collected and the emerging patterns, to determine the number of units in the perimeters and areas. They should realize that the model letter's perimeter is multiplied by the same number as the scale factor used to generate the flip card letter version. Thus, when the scale factor is *n*, the perimeter, *p*, of the model letter becomes *np* for the perimeter of the flip card letter. And, when the scale factor is *n*, the area, *a*, of the model letter becomes an^2.

The magnitude of the scale factor affects how the size of the new image formed will be changed. All of the scale factors used in the activities are whole numbers greater than 1. If 1 is used as the scale factor, the new image will be identical in shape and size to the original, that is, congruent to the original shape. When the scale factor is greater than 0 but less than 1, the new image will be similar to the original but smaller.

BON VOYAGE

- Area
- Volume
- Spatial reasoning
- Scale drawing

Getting Ready

What You'll Need

Color Tiles, about 75 per pair

Color Tile grid paper, page 115

Snap Cubes, about 150 per pair

Isometric dot paper, page 118

Activity Master, page 102

Overview

Students use Color Tiles to build rectangles with specific dimensions and then look at the changes in area occurring when the lengths and/or widths are increased or decreased. Then they build rectangular solids from Snap Cubes and determine changes in volume when their dimensions are changed. In this activity, students have the opportunity to:

- relate numerical patterns to visual patterns
- calculate areas and volumes
- discover how changes in dimensions affect areas or volumes
- make predictions about other figures based on patterns occurring in the data collected

Other *Super Source* activities that explore these and related concepts are:

Stadium Flip Cards, page 27

Wholes and Holes, page 38

Puzzle in a Puzzle, page 43

The Activity

On Their Own (Part 1)

Tanya is planning a bon voyage party for her friend Carlos. She wants to make different-sized banners for the party using brightly colored squares of construction paper. Can you help her determine the number of paper squares needed to make each banner?

- Work with a partner. Using Color Tiles to represent the construction paper squares, build a rectangle that has a length of 4 units and width of 2 units for Tanya's first banner.
- Draw your model on the grid paper and record its dimensions and the number of construction paper tiles in its area.
- Make models for banners whose dimensions are based on changes to the original 4-by-2 banner as follows: (Be sure to draw the model and record the dimensions and number of tiles used for each new banner.)

1. 4-by-6
2. 12-by-2
3. 12-by-6
4. 2-by-2
5. 4-by-1
6. 4-by-8
7. 2-by-1
8. 8-by-8
9. 8-by-6
10. 2-by-6

- Look for relationships between the dimensions of the newly formed rectangular banners and their areas. Explain the changes in the dimensions and their affect on the area of the rectangle or the number of tiles needed to build it.
- Think about how you might generalize your findings about the area of a banner that is *n* times longer and *m* times wider. Be ready to share your findings.

Thinking and Sharing

Students may choose to focus on one rectangular banner and its relationship to the original banner, or they may elect to group specific sets of rectangles together based on their widths or lengths, or a combination of both values. Invite students to share their strategies and conclusions.

Use prompts like these to promote class discussion:

- What happened to the number of Color Tile construction paper squares needed when you changed only one of the dimensions? changed both of the dimensions?
- Can these results be applied to other rectangles? How?
- Do the dimensions of the original rectangle have any effect on the change in area when its length and width are halved, doubled, tripled, an so on? Why?

On Their Own (Part 2)

What if... ***Tanya and her friends want to buy Carlos a set of luggage as a bon voyage gift? After comparison shopping, they decide to consider how the dimensions of each piece of luggage affect its volume before making the purchase. What conclusions do you think they may have come to?***

- Work with a partner. Using combinations of Snap Cubes to represent the dimensions of the first piece of luggage, build a 5-by-1-by-4 rectangular solid.
- Draw your model on the isometric dot paper and record its dimensions and the number of cubes in its volume.
- Make models for pieces of luggage whose dimensions are based on changes to the original 5-by-1-by-4 solid as follows: (Be sure to draw the model and record the dimensions and number of Snap Cubes used for each new rectangular solid.)

  1. 5-by-2-by-4
  2. 5-by-1-by-8
  3. 10-by-1-by-8
  4. 10-by-2-by-8
  5. 5-by-1-by-2
  6. 10-by-3-by-2
  7. 10-by-2-by-1
  8. 4-by-1-by-8

- Look for relationships between the dimensions of the newly formed rectangular solids and their volumes. Explain the changes in the dimensions and their affect on the volume of the solid, or the number of cubes needed to build it.
- Think about how you might generalize your findings about the volume of a piece of luggage that is *n* times longer, *m* times wider, and *p* times taller. Be ready to share your findings.

Thinking and Sharing

If students have grasped the ideas from the first activity, it will be easier for them to extend the concepts to apply to three-dimensional solids. Allow ample time for students to build the models and sketch them on the isometric dot paper.

Use prompts like these to promote class discussion:

- What happened to the number of Snap Cubes needed when you changed only one of the dimensions? two of the dimensions? all three of the dimensions?
- Can these results be applied to other rectangular solids? How?
- Do the dimensions of the original rectangular solid have any effect on the change in volume when its length, width, and height are halved, doubled, tripled, and so on? Why?

Write a letter to Tanya explaining how the change in the volume of a piece of luggage is affected by the change(s) to the original dimensions.

Teacher Talk

Where's the Mathematics?

The diagrams below illustrate the original Color Tile banner and the 10 other banners' dimensions and areas. It should be noted that for each set of dimensions, the first number represented the length (horizontally) and the second, the width (vertically).

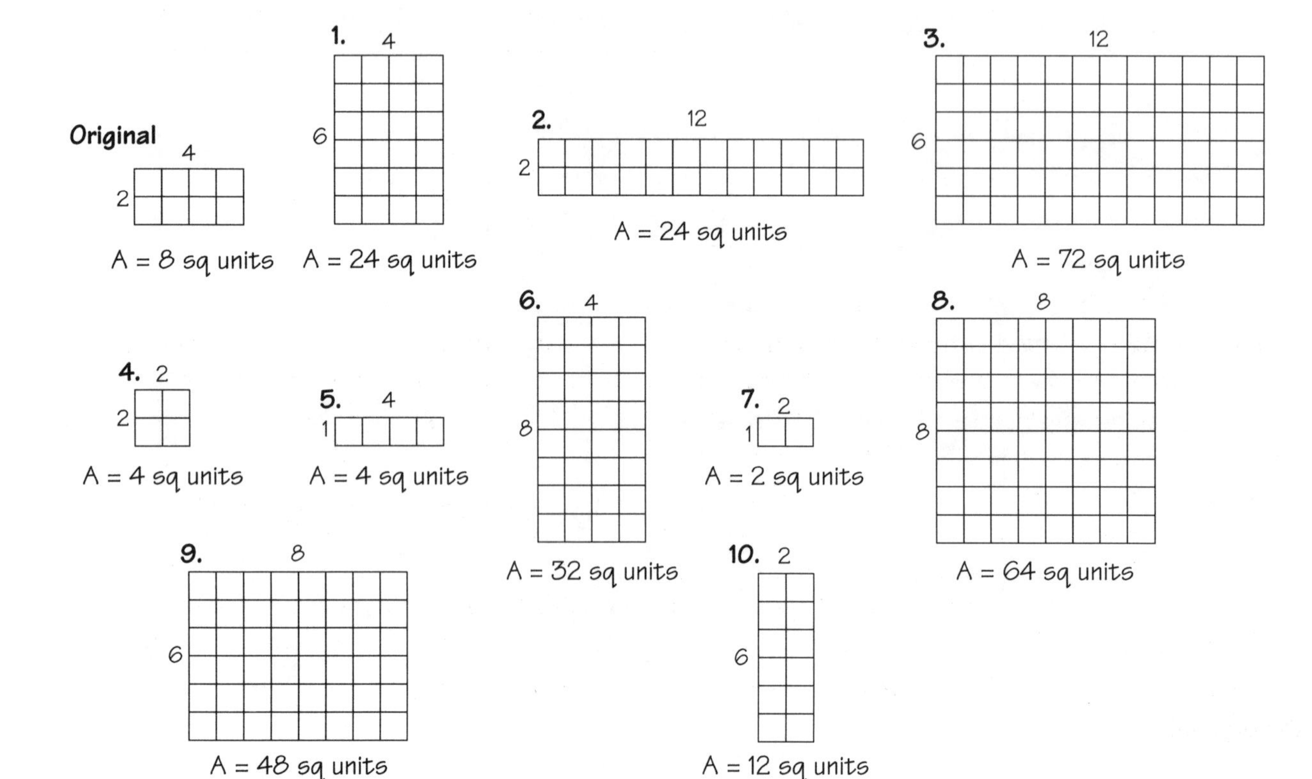

By studying the models where one dimension is changed, students should realize that the change in either the length or width is consistent with the corresponding change in the area.

Banner	Dimensions (L x W)	Area	Change in width	Change in area
Original	4-by-2	8	—	—
1	4-by-6	24	multiplied by 3	multiplied by 3
5	4-by-1	4	multiplied by $\frac{1}{2}$	multiplied by $\frac{1}{2}$
6	4-by-8	32	multiplied by 4	multiplied by 4

Banner	Dimensions (L x W)	Area	Change in length	Change in area
Original	4-by-2	8	—	—
2	12-by-2	24	multiplied by 3	multiplied by 3
4	2-by-2	4	multiplied by $\frac{1}{2}$	multiplied by $\frac{1}{2}$

Banner	Dimensions (L x W)	Area	Change in length/width	Change in area
Original	4-by-2	8	—	—
3	12-by-6	72	length—multiplied by 3 width—multiplied by 3	multiplied by 9
7	2-by-1	2	length—multiplied by $\frac{1}{2}$ width—multiplied by $\frac{1}{2}$	multiplied by $\frac{1}{4}$
8	8-by-8	64	length—multiplied by 2 width—multiplied by 4	multiplied by 8
9	8-by-6	48	length—multiplied by 2 width—multiplied by 3	multiplied by 6
10	2-by-6	12	length—multiplied by $\frac{1}{2}$ width—multiplied by 3	multiplied by $\frac{3}{2}$

If students are able to organize their information in a similar way, they will see that when either the length or width is multiplied by a constant, the new area is changed by the same factor. For example, if the length of the original rectangle is tripled, the area of the original rectangle is also tripled. Or, if the width of the original rectangle is halved, the area of the original rectangle is also halved.

Students will find that when both the length and width of the rectangle are changed, the area of the newly formed rectangle must reflect these two changes. The factors used to change the two dimensions are multiplied together to obtain the change in the area of the original rectangle. For example, if the length of the original rectangle is tripled and the width is doubled, the area of the new rectangle formed will be 2 x 3 or 6 times that of the original. Or, if the length of the original rectangle is halved and the width is multiplied by 5, the area of the new rectangle formed will be $\frac{1}{2}$ x 5 or $\frac{5}{2}$ times that of the original.

In general, if the dimensions of the original rectangle are L by W, its area is LW. When the length is multiplied by n and the width is multiplied by m, the area becomes $(nL)(mW)$ or $(nm)(LW)$, telling us that the new area is (nm) times that of the original rectangle. It can be concluded that the dimensions chosen for the rectangle have no effect on the change in area. Students may notice that the commutative and associative properties have been used in the equation $(nL)(mW) = (nm)(LW)$.

In Part 2, drawing models of the Snap Cube luggage on dot paper requires students to visualize the solid in a three-dimensional position and transfer that concept to the two dimensions of the paper. Graphs of the models are shown below.

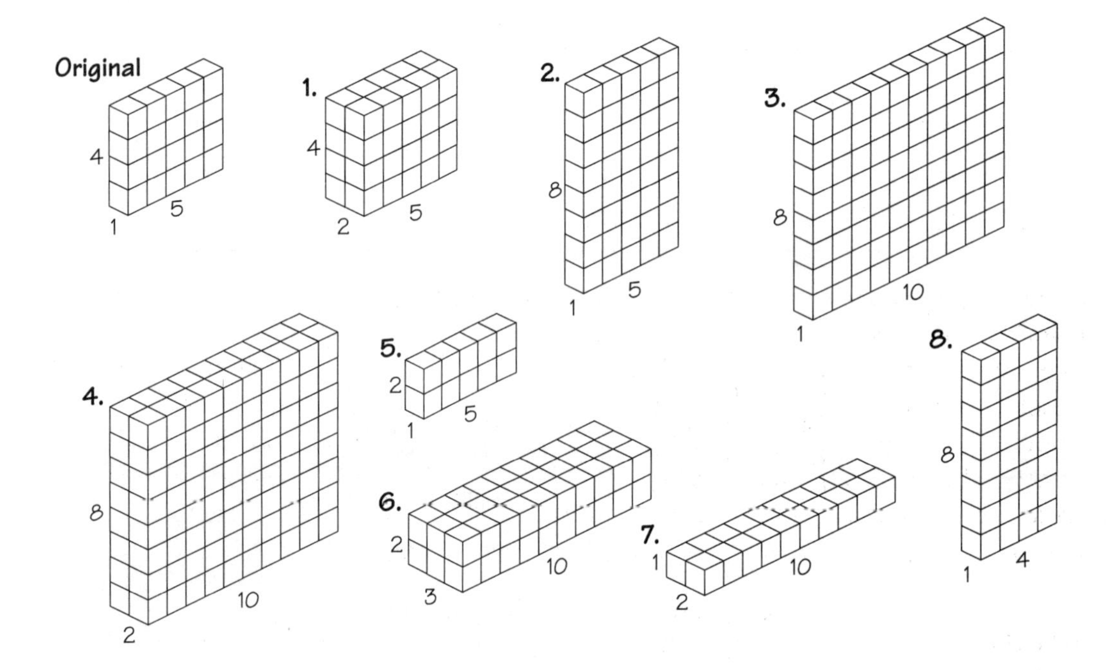

Investigating the changes in volume of the rectangular solids formed using Snap Cubes and arriving at conclusions similar to those above can be seen by considering the example in the second activity. The original 5-by-1-by-4 solid and the other solids formed by changing various dimensions can be seen in the data organized here and on the next page.

Luggage	Dimensions (L x W x H)	Volume	Changes in dimensions	Change in volume
Original	5-by-1-by-4	20	—	—
1	5-by-2-by-4	40	width—multiplied by 2	multiplied by 2
2	5-by-1-by-8	40	height—multiplied by 2	multiplied by 2
5	5-by-1-by-2	10	height—multiplied by $\frac{1}{2}$	multiplied by $\frac{1}{2}$
3	10-by-1-by-8	80	length—multiplied by 2 height—multiplied by 2	multiplied by 4
8	4-by-1-by-8	32	length—multiplied by $\frac{4}{5}$ height—multiplied by 2	multiplied by $\frac{8}{5}$

Luggage	Dimensions (L x W x H)	Volume	Changes in dimensions	Change in volume
Original	5-by-1-by-4	20	—	—
4	10-by-2-by-8	160	length—multiplied by 2 width—multiplied by 2 height—multiplied by 2	multiplied by 8
6	10-by-3-by-2	60	length—multiplied by 2 width—multiplied by 3 height—multiplied by $\frac{1}{2}$	multiplied by 3
7	10-by-2-by-1	20	length—multiplied by 2 width—multiplied by 2 height—multiplied by $\frac{1}{4}$	no change

Students will find that when one or more of the dimensions of the Snap Cube rectangular solid are changed, the volume of the newly formed solid must reflect these changes. The factors used to change the dimensions of the original solid will determine the change in the volume of the new solid. For example, if the one dimension of the original rectangular solid is tripled, a second dimension is doubled, and the third dimension remains the same, the volume of the new rectangular solid formed will be 3 x 2 x 1 or 6 times that of the original. Or, if one dimension of the original rectangular solid is halved, a second dimension is quartered, and the third dimension is multiplied by 5, the volume of the new rectangular solid formed will be $\frac{1}{2} \times \frac{1}{4} \times 5$ or $\frac{5}{8}$ times that of the original.

In general, if the dimensions of the original rectangular solid are L by W by H, its volume is LWH. When the length is multiplied by n, the width is multiplied by m, and the height is multiplied by p, the volume becomes $(nL)(mW)(pH)$ or $(nmp)(LWH)$, telling us that the new volume is (nmp) times that of the original rectangular solid. It can be concluded that the dimensions chosen for the Snap Cube solid have no effect on the change in volume. Again, the commutative and associative properties have been used in the equation $(nL)(mW)(pH) = (nmp)(LWH)$.

WHOLES AND HOLES

- Area
- Pick's theorem

Getting Ready

What You'll Need

Geoboards, 1 per student

Rubber bands

Geodot paper, page 119

Activity Master, page 103

Overview

Students determine the area of a quadrilateral on a Geoboard using Pick's theorem. Then students create and determine the area of a donut-shaped region that is formed by two rubber-band polygons placed on the Geoboard, one inside the other. In this activity, students have the opportunity to:

- apply Pick's theorem to find areas of triangles
- reinforce the concept that the whole is equal to the sum of its parts
- draw models of Geoboard polygons
- test hypotheses about Pick's theorem and polygons
- problem solve to create a "donut" shape having a given area

Other *Super Source* activities that explore these and related concepts are:

Stadium Flip Cards, page 27

Bon Voyage, page 32

Puzzle in a Puzzle, page 43

The Activity

On Their Own (Part 1)

Jamie wants to know if she can find the area of a quadrilateral built on a Geoboard, based on what she already knows about Geoboard triangles and Pick's theorem. Help her explore this question for different quadrilaterals. You will need to know Pick's theorem: Area of a Geoboard triangle = $B/2 + I - 1$, where B represents the number of border pegs and I represents the number of interior pegs.

- Work with a partner. Find the number of units in the large Geoboard square. Record this information.

- Using one rubber band, build a quadrilateral on the Geoboard in which each vertex is a peg on a different side of the Geoboard square's perimeter. Place additional rubber bands on the Geoboard to outline the surrounding triangles. Each side of the quadrilateral corresponds to one side of each triangle.
- On Geodot paper, draw a model of the Geoboard quadrilateral and the surrounding Geoboard triangles.
- Applying Pick's theorem, find the area of each surrounding Geoboard triangle and record the data for the appropriate triangle on the diagram.
- Find the area of the Geoboard quadrilateral using the data you have collected. Be ready to explain the process used.
- Apply Pick's theorem directly to the Geoboard quadrilateral to find its area and compare this to the area previously found.
- Repeat the entire process for either a Geoboard pentagon or hexagon whose vertices are pegs on the perimeter of the Geoboard square. Be ready to discuss your findings.

Thinking and Sharing

Invite students to post diagrams and solutions for their Geoboard quadrilaterals and explain the methods used to find their area. Students should be able to see and verify the concept that the "whole is equal to the sum of its parts" using the Geoboard figures and their area computations. Repeat the process for the Geoboard pentagons and hexagons.

Use prompts like these to promote class discussion:

- How did you choose to place the rubber band to form the original quadrilateral?
- What was the area of the Geoboard square?
- What were the areas of the triangles using Pick's formula?
- How did you use this information in determining the area of the enclosed quadrilateral?
- When you applied Pick's theorem to find the quadrilateral area, how did this result compare to the area you already found?
- Could you use a similar process in determining the area of the pentagons, hexagons, and *n*-gons, in general? Explain.
- Explain the meaning of the phrase, "The whole is equal to the sum of its parts."

On Their Own (Part 2)

What if... ***Jamie is interested in finding the area enclosed between two polygons? She calls these figures "donuts" and wonders if she can use the Geoboard and apply Pick's theorem to these new shapes. Can you help her investigate?***

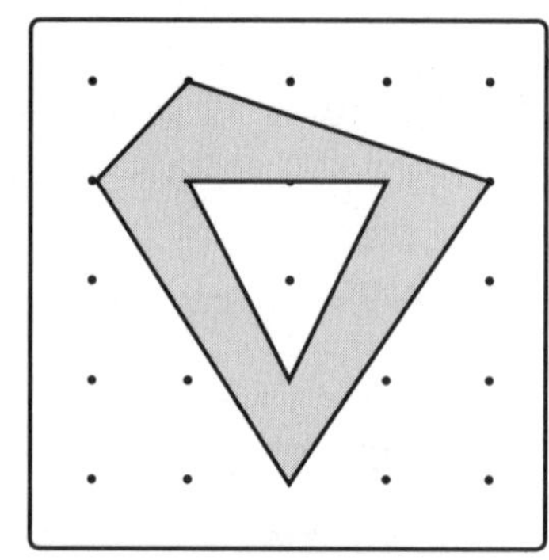

- Using the Geoboard and two rubber bands, create a donut similar to the one shown here. Be sure that the "donut hole" does not have any pegs in common with the donut's outer edge. Work so that the other groups cannot see the donut you are designing.
- Using Pick's theorem, explore ways in which the area of the donut can be determined.
- Draw a model of the Geoboard donut and record the area(s) found on the diagram.
- Exchange the value of the donut area you found with that of another group. Try to build a Geoboard donut with the given area.
- When both groups have finished, check the results against the other group's donut diagram. Discuss any difference and be ready to discuss your findings.

Thinking and Sharing

Ask students how they went about designing their donuts and what happened when they tried to solve the donut problem created by the other group. Invite students to share their designs and ideas.

Use prompts like these to promote class discussion:

- How did you go about creating your donut?
- How is this shape different from the polygons in the first activity?
- How did you find its area?
- How did you go about building the other group's donut?
- Did you and your partner build a donut that was different from the one intended by the other group? If so, why did this happen? Did both donuts satisfy the area requirement?
- Can Pick's theorem be applied to donut shapes? Explain.

Write to Jamie explaining how these methods can be applied to find the areas of polygons, in general, or the area of regions located between two given shapes.

Teacher Talk

Where's the Mathematics?

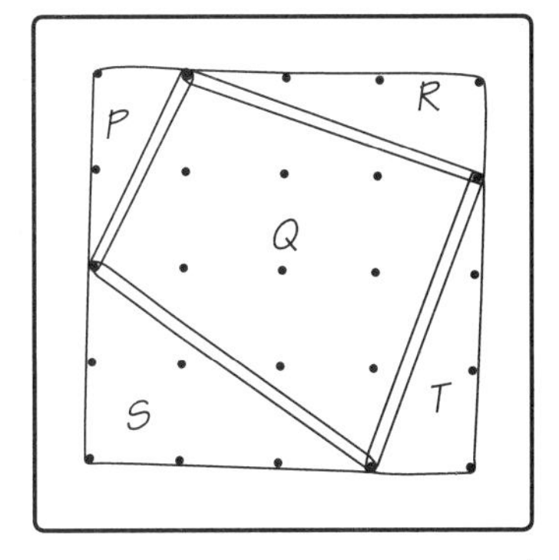

As students present diagrams for finding the areas of the rubber band quadrilaterals, it is important to check that the quadrilaterals have their vertices on perimeter pegs located on different sides of the Geoboard perimeter to insure that the surrounding polygons are triangles. An example is shown here.

When students know Pick's theorem for finding the area of a triangle on the Geoboard, they should realize that by finding the areas of the surrounding triangles and subtracting the sum of their areas from the area of the Geoboard square, they will obtain the area of their quadrilateral. In this example, the area of the Geoboard square is 16 square units. Applying Pick's theorem to the four triangular regions yields the following areas:

Area of triangle $P = \frac{4}{2} + 0 - 1 = 1$ square units

Area of triangle $R = \frac{5}{2} + 0 - 1 = 1\frac{1}{2}$ square units

Area of triangle $S = \frac{6}{2} + 1 - 1 = 3$ square units

Area of triangle $T = \frac{5}{2} + 0 - 1 = 1\frac{1}{2}$ square units

Total of areas $= 7$ square units

Therefore, the area of quadrilateral $Q = 16 - 7 = 9$ square units.

If students apply Pick's theorem directly to the quadrilateral, the area obtained is $\frac{4}{2} + 8 - 1 = 9$ square units, which agrees with the area found above. This result may lead students to conclude that Pick's theorem will work for quadrilaterals as well as for triangles. By sharing the students' solutions with the class, they will discover that any of their quadrilateral areas can be found by using Pick's theorem.

Similar conclusions will also be confirmed for pentagons and hexagons. Pick's theorem will enable students to find their areas directly, rather than by surrounding them with other triangles, and going through the longer subtraction method initially used above.

The concept that "the whole is equal to the sum of its parts" is implied in the methods discussed above. By adding the areas of the polygons that make up a figure, students may be able to find the areas of figures for which there is no available area formula. This method will be demonstrated in the second activity.

One method used to find the area of the donut in Part 2 involves partitioning it into smaller polygons, triangles, quadrilaterals, and so on, for which Pick's theorem is known to work.

If the example shown in the second activity is subdivided into a triangle and two quadrilaterals, as shown here, the individual areas can be calculated and added together to find the area of the donut.

Area of triangle $A = \frac{6}{2} + 0 - 1 = 2$ square units

Area of quadrilateral $B = \frac{4}{2} + 1 - 1 = 2$ square units

Area of quadrilateral $C = \frac{4}{2} + 1 - 1 = 2$ square units

Area of "donut" = Total of areas = 6 square units

Other students, using Pick's theorem, may decide to find the area of the larger outer polygon (donut edge) and subtract the area of the smaller inner polygon (donut hole) to find the area of the donut.

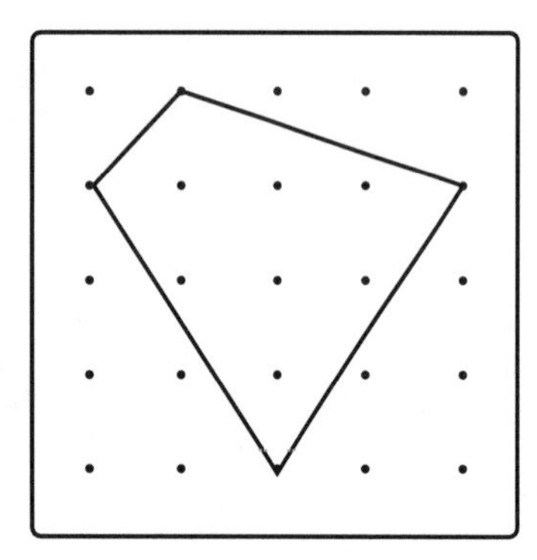

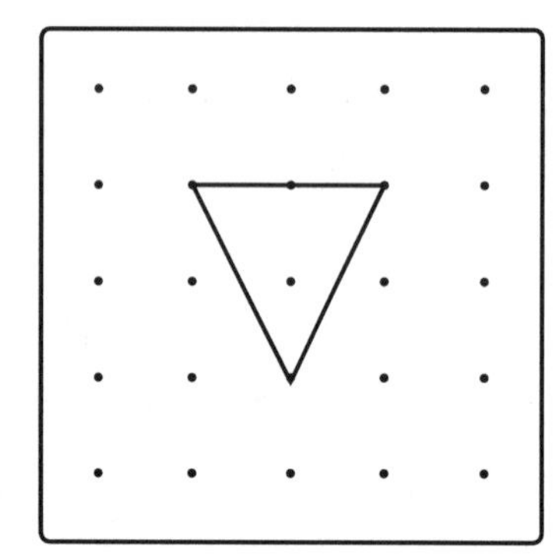

Area of outer polygon = $\frac{4}{2} + 7 - 1 = 8$ square units

Area of inner polygon = $\frac{4}{2} + 1 - 1 = 2$ square units

Area of "donut" = Difference of areas = 6 square units

If students try applying the theorem directly to the donut shape, the area found does not agree with the area found above: $\frac{8}{2} + 2 - 1 = 5$ square units. This missing 1 square unit can be found by looking at the one interior peg from the "donut" hole. It had been counted as an interior peg for the outer polygon and the inner polygon in the first two solutions above, but is not counted as an interior peg for the "donut" hole. Thus, in general, students can conclude that Pick's theorem is appropriate for polygons but not necessarily for other shapes.

PUZZLE IN A PUZZLE

- Area
- Spatial visualization
- Fractions
- Percents

Getting Ready

What You'll Need

Tangrams, 4 sets per group

Pattern Blocks, (no tan or orange) about 60 per group

Tangram paper, page 120

Hexagon silhouette, page 121

Activity Master, page 104

Overview

Students start with a given Tangram piece and build shapes similar to the given piece using a combination of Tangram pieces. They must then determine what relation the area of the new shape has to the area of the original shape. In this activity, students have the opportunity to:

- use spatial reasoning to build shapes
- understand the concept of figures being similar
- calculate areas and fractional parts of areas
- devise strategies for comparing areas

Other *Super Source* activities that explore these and related concepts are:

Stadium Flip Cards, page 27

Bon Voyage, page 32

Wholes and Holes, page 38

The Activity

On Their Own (Part 1)

Sade has designed a set of Tangram challenges called "Puzzle in a Puzzle." Can you solve them?

- Working with a partner, you will need 4 sets of Tangrams to get started. Select the Tangram square from the first Tangram set; this will be your model. Then build three squares of different sizes that are similar to the model but that satisfy the following conditions:
  - the *small* square uses 3 pieces from the second Tangram set
  - the *medium* square uses 5 pieces from the third Tangram set
  - the *large* square uses all 7 pieces from the fourth Tangram set
- Assuming that the side length of the original Tangram model has a side length of 1 unit, find the side lengths of the *small, medium,* and *large* squares. Find and record the area of the original Tangram model and determine the areas of the *small, medium,* and *large* squares.

- Look for relationships between the set of squares, their side lengths, and areas.
- Now, begin a similar process by selecting the smallest Tangram triangle from the first Tangram set. Use it as the model for building three similar triangles of different sizes satisfying the following conditions:
  - the *small* triangle uses 3 pieces from the second Tangram set
  - the *medium* triangle uses 5 pieces from the third Tangram set
  - the *large* triangle uses all 7 pieces from the fourth Tangram set
- Assuming that the area of the large triangle (using all 7 pieces) has an area of 1 square unit, find and record the areas of the *model*, *small*, and *medium* triangles.
- Look for relationships between the set of triangles and their areas. Be ready to discuss your findings.

Thinking and Sharing

Invite students to post a square that demonstrates each condition. Then have students post their triangles. Discuss the relationships between the models and their similar replicas.

Use prompts like these to promote class discussion:

- What was easy about building the small, medium, and large shapes? What was hard?
- How did you find the side length and area of each size version of the model?
- What relationships did you discover about the shapes and their areas?
- Did you discover any patterns? Explain.
- What does it mean for figures or shapes to be "similar"?

On Their Own (Part 2)

What if... ***Sade created another "Puzzle in a Puzzle" challenge, but this time she used Pattern Blocks instead of Tangrams? Can you solve this "Puzzle in a Puzzle"?***

- Working with a partner, use Pattern Block pieces to completely fill in the outline on the Hexagon Silhouette page. Follow the rules stated below:
  - only yellow, red, blue, and green pieces may be used
  - there must be a different number of each shape used to complete the design
- Trace the outline of each Pattern Block shape on the worksheet, labeling each piece's color.
- If the area of a blue rhombus is assumed to be 1 square unit, find the total area of the hexagon. Record your findings.

- Find the total area made up by Pattern Blocks of the same color based on the blue rhombus representing 1 square unit. Record your findings.
- Determine the fractional part or percentage of the whole area represented by the area of all Pattern Blocks of the same color.
- Be ready to share your methods and results.

Thinking and Sharing

Students will complete the hexagon silhouette using different combinations of Pattern Blocks. Their results on the area or percent occupied by blocks of the same color may therefore vary from group to group. However, by adding up the fractional parts of the whole or the percentages represented by the different colors, each group should come up with the total of 1 or 100%.

Use prompts like these to promote class discussion:

- How did you choose to fill in the silhouette? What was easy? What was hard?
- How many of each different Pattern Block color did you use in the hexagon outline?
- What was the area of the hexagon silhouette? How did you find it?
- What is the total area made up of Pattern Blocks of the same color? How did you find it?
- How did you find the fractional part or the percentage of the entire hexagon's area represented by the total area of each color?
- How can you check your results?

Write a letter to Sade describing how you might construct a "Puzzle in a Puzzle" for her to solve. Explain or list the steps she should follow to solve your challenge.

Teacher Talk

Where's the Mathematics?

When students start building the larger versions of the model square, they may have trouble deciding which pieces to choose. Working with a partner may help some students who have difficulty with spatial reasoning and visualization of geometric relationships. Other students may choose to build the *large* square first because they know that the entire set of 7 Tangram pieces is used and then try to build the *small* and *medium* squares.

The diagrams below show the model square and the three different-sized versions.

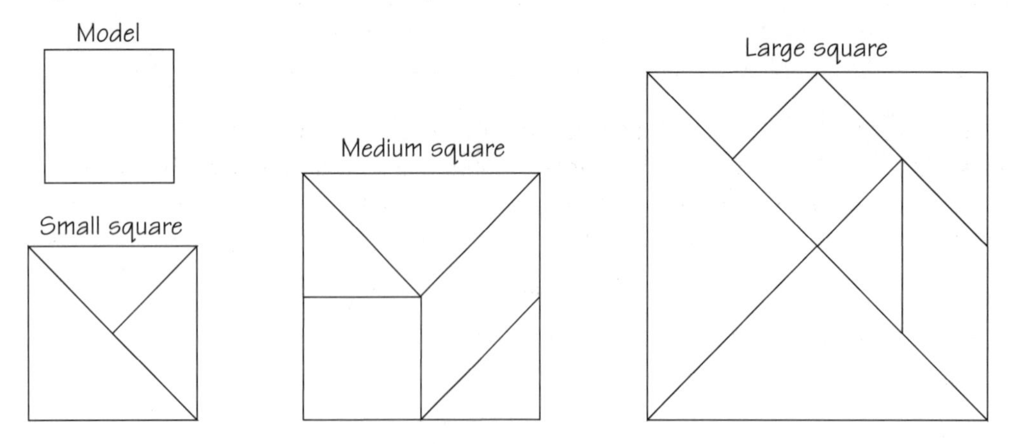

	Model	Small square	Medium square	Large square
Area	1 sq unit	2 sq units	4 sq units	8 sq units
Side length	1 unit	$\sqrt{2}$ units	2 units	$\sqrt{8}$ units

One method used to find the areas of the medium and large squares is based on the number of Tangram triangles (smallest size) into which each square can be partitioned. The model square can be thought of as 2 triangles, the *small* square as 4 triangles, the *medium* square as 8 triangles, and the *large* square as 16 triangles. By simplifying the ratio 2:4:8:16 into the ratio 1:2:4:8, where 1 square unit represents the area of the model square, the areas of the other three figures can be determined. These sets of four terms form a geometric sequence where the ratio between two consecutive terms is 2.

By applying the area formula for a square, $A = s^2$, to each of the known areas above, students can find the side lengths of $1, \sqrt{2}, 2, \sqrt{8}$ or $2\sqrt{2}$ for each of the four figures respectively. For example, for the small square:

$$A = s^2 \qquad 2 = s^2 \qquad \sqrt{2} = s$$

This set of four terms forms a geometric sequence where the ratio between two consecutive terms is $\sqrt{2}$.

The triangle "Puzzle in a Puzzle" presents a slightly more challenging problem. The triangles are all similar to the original isosceles right triangle model. This means that all of the corresponding angles of the new versions are congruent (equal in measure) to those of the model and the corresponding sides are all in proportion.

The diagrams below show the model triangle and the three different-sized versions.

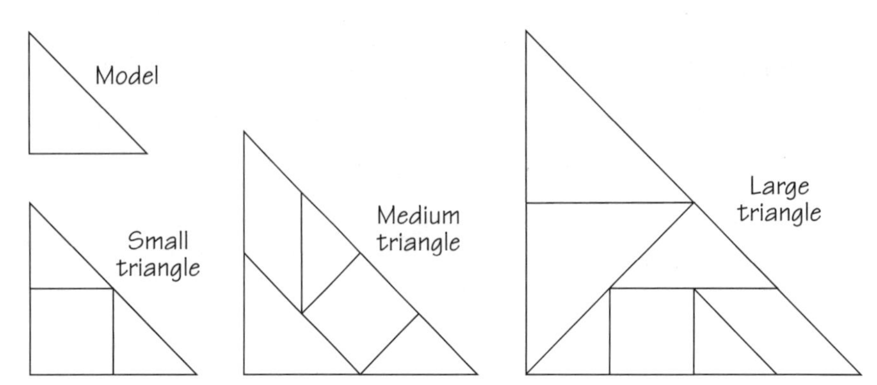

	Large triangle	Medium triangle	Small triangle	Model
Area	1 sq unit	$\frac{1}{2}$ sq unit	$\frac{1}{4}$ sq unit	$\frac{1}{16}$ sq unit

One method used to find the area of the smaller version triangles is based on the number of Tangram triangles (smallest size) into which the largest triangle can be partitioned. The large triangle can be thought of as 16 small Tangram triangles, the medium triangle as 8 triangles, the small triangle as 4 triangles, and the model triangle as 1 triangle. By simplifying the ratio 16:8:4:1 into the ratio $\frac{16}{16}:\frac{8}{16}:\frac{4}{16}:\frac{1}{16}$ or $1:\frac{1}{2}:\frac{1}{4}:\frac{1}{16}$, where 1 square unit represents the area of the large triangle, the areas of the other three figures can be determined. This set of four terms begins a power sequence of the form: $\frac{1}{2}\,n - 1$ where $n = 1, 2, 3, 4$.

The second activity provides students the opportunity to create different "fillings" for the hexagonal silhouette. (A different number of each Pattern Block color is required in the rules so that each color will constitute a different fractional part of the whole.) In the example below, 10 green blocks, 7 blue blocks, 4 red blocks, and 3 yellow blocks have been used.

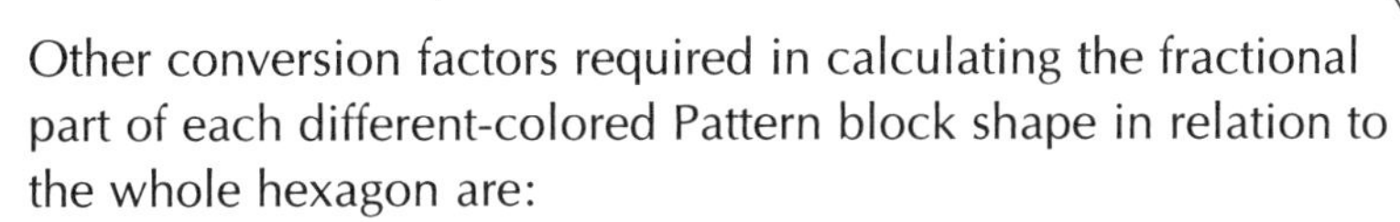

To determine the area of the hexagon in terms of the blue rhombus, whose area represents 1 square unit, students may choose to look at the number of green blocks that would exclusively cover the hexagon and convert that number into the appropriate number of blue rhombi. Fifty-four green Pattern Blocks are needed to cover the hexagon silhouette or the equivalent of 27 blue rhombi.

Other conversion factors required in calculating the fractional part of each different-colored Pattern block shape in relation to the whole hexagon are:

1 green block = $\frac{1}{2}$ of the blue rhombus

1 red block = $\frac{3}{2}$ of the blue rhombus

1 yellow block = 3 blue rhombi

When these conversion factors are applied to the Pattern Blocks making up the hexagon design above, it is found that the 10 green blocks are equivalent in area to 5 blue blocks, 4 red blocks are equivalent to 6 blue blocks, and 3 yellow blocks are equivalent to 9 blue blocks.

Thus, the fractional part of the area of the large hexagon occupied by the area of each color Pattern Block is:

green blocks = $\frac{5}{27}$ of the hexagon's area or 18.52%

blue blocks = $\frac{7}{27}$ of the hexagon's area or 25.93%

red blocks = $\frac{6}{27}$ of the hexagon's area or 22.22%

yellow blocks = $\frac{9}{27}$ of the hexagon's area or 33.33%

To check: Total = $\frac{27}{27}$ = 1 hexagon's area or 100%

Investigating Volume and Surface Area

The lessons in this cluster reinforce comprehension of volume, surface area, and nets, practice in three-dimensional modeling, and provide creative exploration. The third lesson should be done after the first and/or second lessons are completed.

1. Storage Boxes (*Investigating volumes and surface areas of rectangular solids*)

In this lesson, students explore the range of possible rectangular solids obtained by arranging a given set of Cuisenaire Rods. As they search to find all possible solids, students discover attributes of shapes that affect surface area.

The activity and discussion questions help students to clarify the concepts of volume and surface area. They also prompt students to consider the relationship between these two measurements. In *For Their Portfolio* (page 51) students may create a net of a cardboard storage box.

2. Cube Sculptures (*Investigating surface areas of solids having the same volume*)

In their search to find as many arrangements of a given set of Snap Cubes as possible, students devise strategies for generating different surface areas. Students will be challenged when comparing the isometric models of their shapes to others who have drawn the same shapes from different perspectives. They may realize that shapes that look totally different may have equal surface areas and that regardless of the arrangement, the number of visible faces is always even.

On Their Own Part 2 has students visualize a prism that will contain each arrangement and then determine its volume. Students may be surprised to find that the prisms needed to enclose the structure with the smallest surface area and the structure with the largest surface area have the same volume.

3. Wrapping Paper (*Investigating surface areas and nets of solids*)

On Their Own Part 1 challenges students to attach a given number of Snap Cubes to form as many different rectangular solids as possible, find their surface areas, and design nets that could be folded to cover the surfaces of these solids. Lively class discussion may result when students represent identical solids on isometric grid paper from different perspectives and generate different nets to cover them.

On Their Own Part 2 extends the investigation for arrangements of Snap Cubes that are not rectangular solids. *Where's the Mathematics?* (page 60) provides examples of different methods students with varying amounts of prior knowledge may use to solve the problem of finding surface areas.

STORAGE BOXES

- *Volume*
- *Surface area*
- *Spatial visualization*

Getting Ready

What You'll Need

Red Cuisenaire Rods, 32 per pair

Metric rulers

Isometric dot paper, page 118

Activity Master, page 105

Overview

Students investigate volume and surface area by modeling arrangements of shoe boxes using Cuisenaire Rods. In this activity, students have the opportunity to:

- reinforce the concepts of volume and surface area
- create 3-dimensional scale models
- experiment with ways to conserve surface area
- strengthen spatial visualization

Other *Super Source* activities that explore these and related concepts are:

Cube Sculptures, page 53

Wrapping Paper, page 57

The Activity

On Their Own (Part 1)

Kathryne takes care of her shoes by keeping them in their original shoe boxes. She wants to find one large storage box that will hold 8 of her shoe boxes. Can you help Kathryne determine the dimensions of the storage boxes that would work?

- Work with a partner. Use red Cuisenaire Rods to represent the shoe boxes.
- Arrange 8 shoe boxes so that they could fit into a rectangular prism-shaped storage box. The box should be exactly the right size to hold the 8 shoe boxes with no extra space left over. Find as many different arrangements as possible.
- Record your models on isometric dot paper. Measure and record the dimensions and volumes of each model.
- Determine the *actual* dimensions of each of Kathryne's shoe boxes if 1 centimeter in your model represents 15 centimeters for the *actual* shoe boxes. Then calculate the actual dimensions and volumes of the storage boxes that you modeled.
- Be ready to discuss your findings.

Thinking and Sharing

Ask different pairs of students to share one of their models and record its measurements and the measurements of the *actual* storage box on the board. Continue until all models have been displayed and discussed.

Use prompts like these to promote class discussion:

- How did you go about finding different arrangements of the shoe boxes?
- How did you know that you had found all of the possible arrangements?
- How did you find the dimensions and volumes of your models?
- How did you calculate the dimensions and volumes of the actual storage boxes?
- What did your findings reveal about the dimensions and volumes of the possible storage boxes?
- Which storage box do you think would be the most practical for storing the shoe boxes? Explain.
- What patterns do you notice in the class chart?

On Their Own (Part 2)

What if... *Kathryne decides to make her own storage box from sheets of plywood? What is the least amount of plywood she would need to create the box? What is the most?*

- Using your models from Part 1, determine the amount of plywood needed to construct each possible storage box. Record your measurements (in square centimeters) near your drawings.
- Determine which arrangement would need the least amount of plywood and which would need the most.
- Now use your observations to determine the least and greatest amounts of plywood needed to construct a storage box that would hold 12 shoe boxes.
- Try to come up with a general rule that could be used to predict what kinds of box arrangements will use the least amount of plywood.
- Be ready to explain your methods and discuss your findings.

Thinking and Sharing

Invite pairs to discuss their results. Have them add the surface area measurements to those already recorded on the board (from Part 1).

Use prompts like these to promote class discussion:

- How did you determine the amount of plywood needed to build each of your storage boxes?
- What did you find to be the least and greatest amounts of plywood needed?

- Which shoe-box arrangement uses the least amount of plywood? the greatest? How do these arrangements compare?
- What did you determine to be the least and greatest amounts of plywood needed to construct the storage box that would hold 12 shoe boxes? Explain your thinking.
- Were you able to find a general rule for minimizing the amount of plywood needed? How could you test that rule?
- How are the dimensions, volumes, and surface areas of your storage boxes related?

Imagine that the storage container is to be made of cardboard instead of plywood. Using your results from Part 2, make a scale drawing of a one-piece pattern that can be cut out and folded to form the storage box needing the least amount of cardboard. Use the same scale as you used for your models. Label your pattern with the actual measurements that would be needed to construct the storage box.

Teacher Talk

Where's the Mathematics?

Using Cuisenaire Rods to model the shoe boxes makes it easy for students to make and compare an assortment of possible storage arrangements. They can also come to recognize that it is possible to make a variety of different-looking structures that all have the same volume.

Students should find that each red Cuisenaire Rod measures 2 cm x 1 cm x 1 cm. Using the given scale factor, they can calculate the actual dimensions of the shoe boxes, which are 30 cm x 15 cm x 15 cm.

There are a variety of ways in which the eight shoe boxes can be arranged to fit inside a rectangular prism-shaped storage box. Six possible arrangements are shown here.

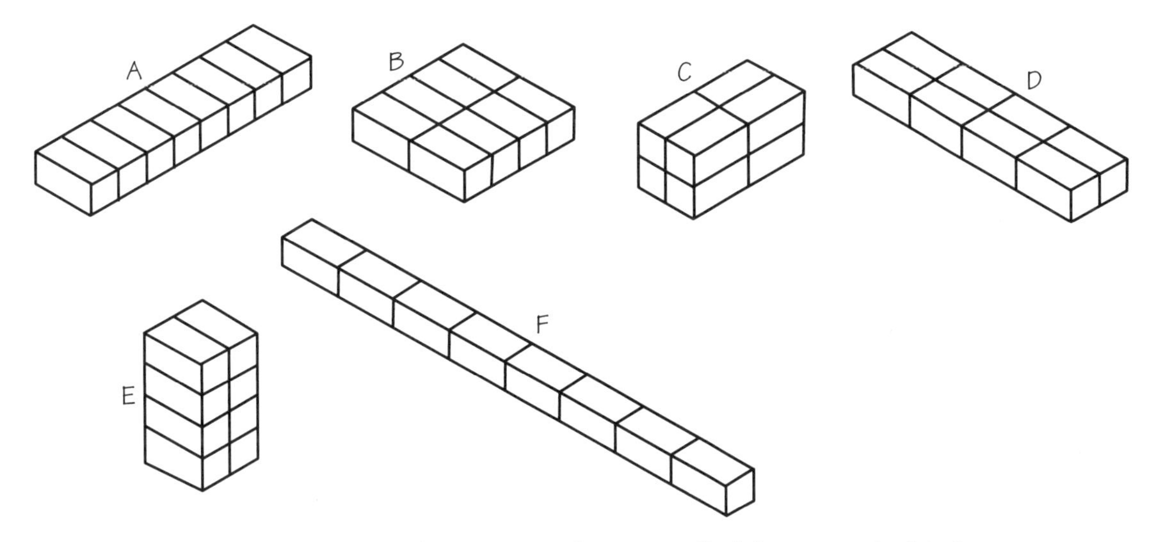

Each of the models has a volume of 16 cm^3. Students may find that several of their arrangements have the same dimensions. For example, of those pictured, A and D both measure 8 cm x 2 cm x 1 cm, and C and E both measure 4 cm x 2 cm x 2 cm. The dimensions of some of the students' actual storage boxes may also be the same.

No matter how they arrange their shoe boxes, students should find that only four different-sized storage boxes are possible: a 240 cm x 15 cm x 15 cm storage box, a 120 cm x 30 cm x 15 cm storage box, a 60 cm x 60 cm x 15 cm storage box, and a 60 cm x 30 cm x 30 cm storage box. Each of these has a volume of 54,000 cm^3. Students can calculate the volumes of their storage boxes by first multiplying the dimensions of their models by 15 (the scale factor) and then finding their product (Volume = length x width x height), or by finding the volume of one shoe box (6750 cm^3) and then multiplying by 8 (the number of shoe boxes). Still another method would be to find the volume of a model and multiply by 15 (length) x 15 (width) x 15 (height), or 3375.

To determine the amount of plywood needed to make the storage boxes, students need to calculate the surface area of each of their arrangements. The surface areas can be determined by finding the total of the areas of the six faces (front, back, left, right, top, and bottom) of their arrangements. The surface areas of the different-shaped storage boxes are given in this table.

Storage Boxes	Area of each side in cm^2						Total Surface Area
	Front	Back	Left	Right	Top	Bottom	
A	30 x 15 450	30 x 15 450	120 x 15 1800	120 x 15 1800	120 x 30 3600	120 x 30 3600	11,700 cm^2
B	60 x 15 900	60 x 15 900	60 x 15 900	60 x 15 900	60 x 60 3600	60 x 60 3600	10,800 cm^2
C	30 x 30 900	30 x 30 900	60 x 30 1800	60 x 30 1800	30 x 60 1800	30 x 60 1800	9,000 cm^2
D	120 x 15 1800	120 x 15 1800	30 x 15 450	30 x 15 450	120 x 30 3600	120 x 30 3600	11,700 cm^2
E	30 x 60 1800	30 x 60 1800	30 x 60 1800	30 x 60 1800	30 x 30 900	30 x 30 900	9,000 cm^2
F	240 x 15 3600	240 x 15 3600	15 x 15 225	15 x 15 225	240 x 15 3600	240 x 15 3600	14,850 cm^2

The arrangement requiring the least amount of plywood is the one represented by the 4 cm x 2 cm x 2 cm model (C and E); the one requiring the most is the one represented by the 16 cm x 1 cm x 1 cm model (F). Students should recognize that the smallest surface area is created by making the rectangular solid as compact as possible. The more elongated and less compact the solid, the greater the surface area, as evidenced by the 14,850 cm^2 of plywood needed to construct the storage box for the arrangement pictured in F. Using this understanding, students can work with their Cuisenaire Rods to create the most compact and most elongated models of arrangements of 12 shoe boxes to determine the smallest and greatest amounts of plywood needed to construct these storage boxes. A compact, 4 cm x 3 cm x 2 cm model would represent a storage box that would require 11,700 cm^2 of plywood, while an arrangement in which the 12 boxes were placed end to end would require 22,050 cm^2 of plywood. Some students may generalize this rule of compactness by stating that the three dimensions should be numerically as close as possible.

CUBE SCULPTURES

- Surface area
- Volume
- Spatial visualization

Getting Ready

What You'll Need

Snap Cubes, about 70 per pair

Isometric dot paper, page 118

Activity Master, page 106

Overview

Students search to find all possible surface areas that can be created by building structures made from 16 Snap Cubes. They then investigate the dimensions and volumes of rectangular prisms that could be used to enclose their structures. In this activity, students have the opportunity to:

- find the surface area of a variety of structures
- discover that structures with the same volume may have different surface areas
- discover that different-looking structures that have the same volume may have the same surface area
- recognize that elongated structures have a greater surface area than compact structures that have the same volume
- determine the overall dimensions of an irregular-shaped structure

Other *Super Source* activities that explore these and related concepts are:

Storage Boxes, page 49

Wrapping Paper, page 57

The Activity

On Their Own (Part 1)

The students in the eighth grade art class are using recyclable materials to build cubes which they will join together to make sculptures for the school courtyard. Each sculpture will be made from 16 cubes. The visible faces of each cube will be painted in different colors. How many different colors are needed?

- Work with a partner. Use Snap Cubes to design models of several different sculptures, each containing 16 cubes.
- Determine the number of colors that would be needed to paint each sculpture. Remember, each visible face must be a different color.
- Record your sculptures on isometric dot paper. Record the volume and surface area of each sculpture using the edge of one cube as the unit of measure.

- Now try to make models of sculptures that will require different numbers of colors from those you recorded. When you find one, record it as you did before.
- Continue until you think you've modeled at least one sculpture for every possible number of visible faces. Be ready to discuss your findings.

Thinking and Sharing

Ask students to reconstruct the sculptures they built requiring the smallest number of different paint colors. Do the same for their sculptures requiring the greatest number of different paint colors. Then have students help you list the surface areas (from smallest to greatest) of the different sculptures they built.

Use prompts like these to promote class discussion:

- What did you notice as you built models of different sculptures?
- How do the volumes of your sculptures compare?
- How do the surface areas of your sculptures compare?
- How did you go about building models of sculptures that would require different numbers of colors?
- Did you notice any patterns as you worked? Were the patterns helpful in building new sculptures? If so, explain.

On Their Own (Part 2)

What if... ***the students decide that each sculpture is to be enclosed in a clear plastic rectangular prism that will protect it from the weather. What size prisms will they need to construct?***

- Reconstruct one of your sculpture models. Imagine enclosing it in the smallest possible rectangular prism that could hold it.
- Determine the dimensions and volume of the prism. Record these measurements near the drawing of your sculpture.
- Reconstruct each of your other models and determine the dimensions and volumes of the prisms they would require. Record your findings.
- Compare your different models and the dimensions of their enclosing prisms. Be ready to discuss your observations.

Thinking and Sharing

Ask students to recreate several sculptures that required rectangular prisms with different dimensions. Have students display their sculptures, discuss how they determined the size of the enclosing prisms, and record the dimensions and corresponding volumes on the chalkboard.

Use prompts like these to promote class discussion:

- How did you go about determining the size of the rectangular prisms?
- Were some prisms easier to visualize than others? If so, which ones, and why?
- Were some of the prisms harder to visualize than others? If so, which ones and why?
- What was the size of the smallest prism you needed? How did this prism compare in volume to the sculpture it would enclose?
- What was the size of the greatest prism you needed? How did this prism compare in volume to the sculpture it would enclose?
- What other observations did you make about your models and their prisms?

Suppose you were building sculptures made from 20 cubes. Describe how you would construct the models having the smallest and greatest possible surface areas. Explain how you know that no other sculpture could have a smaller (or greater) surface area than the ones you described.

Teacher Talk

Where's the Mathematics?

It is often the case that students believe that shapes with the same volume have the same surface area. Many students also believe that shapes that look totally different from each other must have different surface areas. As they investigate the problem set forth in Part 1 of this activity, students will be able to discover that both of these ideas are erroneous.

There are 14 different possible surface areas of structures that can be created using 16 Snap Cubes. They range in area from 40 square units to 66 square units. Some examples of structures that students might build and record are shown below with their respective surface areas.

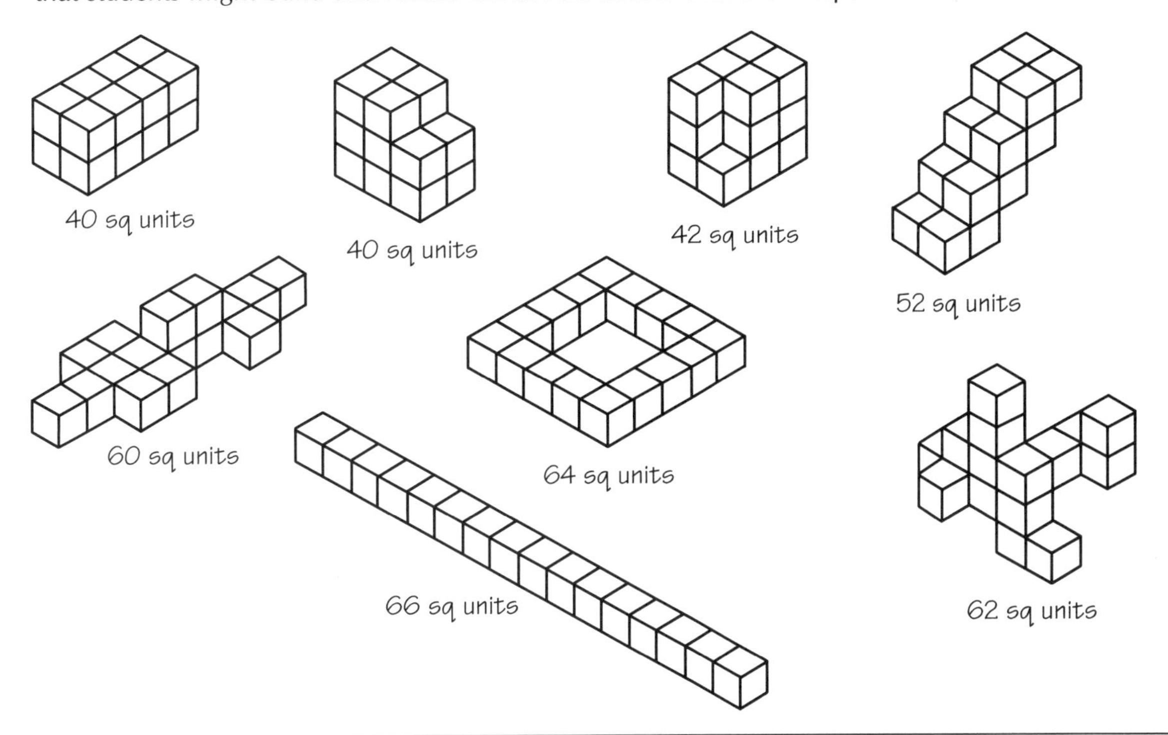

Students may notice that although all of their structures have the same volume (16 cubic units), the structures that are more compact have smaller surface areas than those that are elongated. They may recognize that this is due to the fact that in compact shapes, more cube faces are "hidden" inside the structure and, therefore, do not contribute to the surface area. To build new structures requiring different numbers of paint colors, students may take some of their more compact models, and, by rearranging some of the cubes, "uncover" some of the hidden faces, thereby increasing the surface area.

Students may also notice that the number of different colors needed to paint each of their structures is always an even number. In fact, for every even number from 40 to 66, it is possible to build at least one structure requiring that number of different paint colors. Students may want to investigate why the structures always require an even number of paint colors. As they examine their structures, they may discover that the number of cube faces that are exposed to the front of the structure is equal to the number exposed to the back; the number of cube faces that are exposed to the left of the structure is equal to the number exposed to right; and the number of cube faces that are exposed to the top of the structure is equal to the number exposed to the bottom. Since these numbers occur in pairs, their sum (the total surface area) will always be an even number.

In Part 2, students need to use spatial visualization to picture the smallest rectangular prism that will contain each of their structures. In doing this, they are finding the overall dimensions of their models. Students should recognize that their prisms will need to accommodate the longest row of cubes from each of the three dimensions of their structures. The volume of each prism can then be calculated by multiplying these three measurements together.

One particularly interesting result is that the prisms needed to enclose the structure with smallest surface area (the 2 x 2 x 4 prism) and the structure with greatest surface area (the 1 x 1 x 16 prism) have the same volume (16 cubic units). These are the smallest prisms that will hold a 16-cube structure. Students may notice that in these cases, there is no empty space inside the enclosure once the structure is placed inside. Therefore, the volumes of these enclosures are equal to the volumes of the structures themselves. Larger enclosing prisms will be needed for the other structures, all of which will contain empty space once the structure is placed inside. The structure below would require the largest enclosure, a 6 x 6 x 6 prism, having a volume of 216 cubic units, 200 of which will be empty space once the structure is placed inside.

WRAPPING PAPER

- Nets
- Surface area
- Spatial visualization

Getting Ready

What You'll Need

Snap Cubes, 72 per pair

Rulers

Scissors

Isometric dot paper, page 118

Snap Cube grid paper, page 122

Activity Master, page 107

Overview

Students use Snap Cubes to investigate surface areas and nets of rectangular prisms and irregular-shaped solids. In this activity, students have the opportunity to:

- discover that shapes with the same volume can have different surface areas
- explore how the dimensions of a solid affect its volume, surface area, and net
- learn how to design a net
- use spatial visualization and reasoning to work back and forth between three dimensions and two dimensions

Other *Super Source* activities that explore these and related concepts are:

Storage Boxes, page 49

Cube Sculptures, page 53

The Activity

On Their Own (Part 1)

A candy company packages its candy in individual boxes, which are then wrapped in larger packages containing two dozen small boxes. If the larger packages must be rectangular solids, how many different-sized packages are possible and what patterns can be used to construct them?

- Work with a partner. Imagine that a Snap Cube represents a box that contains one piece of candy. Build three different-sized packages, each containing two dozen of the small candy boxes. Remember that your packages must be in the shape of rectangular solids.
- Find the volume and surface area of each of your packages. Use the length of an edge of one cube as the unit of measure.
- Record each of your packages and its volume and surface area on isometric dot paper.

- Design a one-piece pattern that could be folded up to form the wrapping paper for one of your packages. (You may want to use Snap-Cube grid paper as a tracing guide.) Use dashed lines to indicate where your pattern should be folded once it is cut out. There should not be any overlapping paper.
- Make a copy of your pattern. Cut it out and fold it around your solid to check that it works. If it doesn't, modify your pattern until it does.
- Label your pattern, indicating the length of each side in numbers of units. Record the total number of square units of wrapping paper needed to make your pattern.
- Repeat the process for your other packages. Organize and compare the measurements of each of your packages and be ready to discuss your findings.

Thinking and Sharing

Invite a pair of students to share one of their Snap Cube structures and its one-piece pattern. Have them describe the methods they used for constructing their pattern. Ask another pair to present a different-shaped package or a different pattern. Continue until each pair has presented one of its structures and its pattern. Then have the students help you create a class chart listing the dimensions, volumes, and surface areas of the six different-sized packages that are possible.

Use prompts like these to promote class discussion:

- How did you determine the volumes and surface areas of your packages?
- Do you think we have found all possible packages for the candy boxes? How do you know?
- How did you go about designing the one-piece patterns for your packages?
- How did you determine the amounts of wrapping paper needed for your patterns?
- How do the amounts of wrapping paper needed for the different-sized packages compare?
- What patterns do you notice in the class chart?
- How are the various measurements that you found related to the dimensions of each package?

Tell students that the one-piece patterns they designed are called nets. Explain that nets can be made for all polyhedra, not just for rectangular solids, and that there can be many different nets for the same polyhedron.

On Their Own (Part 2)

What if... ***the company decides to try packaging their candy boxes in packages that are not rectangular solids? What shapes might they consider and what nets can be used to construct them?***

- Using Snap Cubes, work with your partner to design a package that will hold two dozen candy boxes and is not a rectangular solid. Record your package on isometric dot paper.

- Determine the volume and surface area of your package.
- Design a net for your package, using dashed lines to indicate the fold lines. Then make a copy of it, cut it out, and see if it works. If it doesn't, modify it until it does.
- Label your net, indicating the length of each side in numbers of units. Also record the total number of square units of wrapping paper needed to make your net.
- Now make two different nets that could be used to wrap the same package. Again, make copies, cut them out, try them, and modify them if necessary. Label them as you did the first net. Be ready to discuss your observations.

Thinking and Sharing

Have students share their Snap Cube structures and nets. Ask them to discuss how they went about designing the nets for their packages.

Use prompts like these to promote class discussion:

- How did you design your package?
- Was any package difficult to draw using isometric paper? If so, why?
- How did you determine the volume and surface area of your package?
- How did you go about designing your nets? What was difficult about designing them?
- How are your nets alike? How are they different?
- What are some general instructions you might give to someone who needs to design a net for an irregular-shaped solid?

Suppose you know the length, width, and height of a rectangular solid, but you do not have a model. Explain how you would design a net for the solid, and how you would determine how much paper would be needed to make the net.

Teacher Talk

Where's the Mathematics?

In this lesson, students are presented with activities that enable them to explore how the volume and surface area of a solid are related to its dimensions. Students with limited prior exposure to measurement of three-dimensional shapes may be surprised to discover that shapes with the same volume may have different surface areas. In Part 1, students come to see that solids with the same general shape (rectangular solid, in this case) and the same volume may have different surface areas. Their work should reveal that although the products of the dimensions of all of their packages are the same, the different sets of dimensions (reflecting the number of exposed faces of the individual cubes) produce different surface areas.

There are six different-sized packages that can be constructed to hold 24 candy boxes. Any arrangement not pictured below is a different orientation of one of the six shown. Students may point out that no other set of dimensions is possible, as there is no other combination of three whole numbers that multiply to 24.

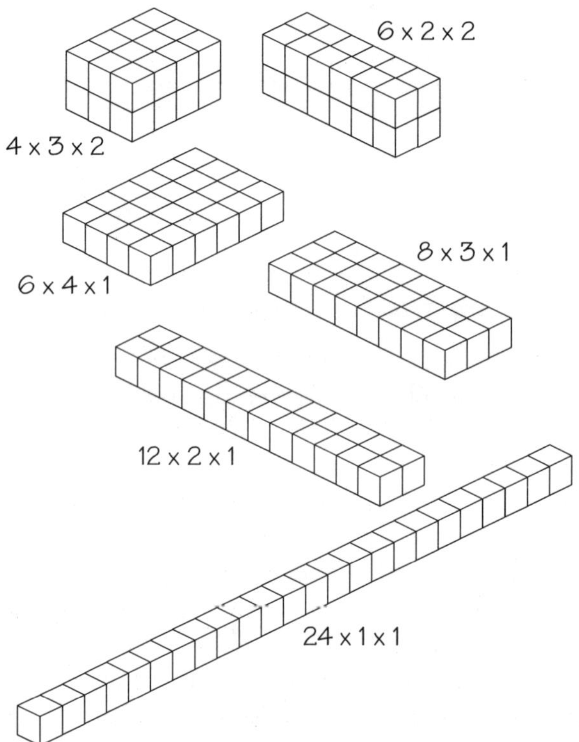

Students should recognize that the volume of each package (24 cubic units) is the product of its dimensions. The surface area of each package can be found by counting the number of exposed cube faces, or by calculating the area of each face of the solid using the formula for finding area of a rectangle (Area = length x width), and then adding the areas together. Students should find that the surface areas range from 52 square units to 98 square units, with the more compact solids having smaller surface areas than those that are more elongated. They may also notice that the surface areas are all even numbers of square units. This can be explained by the fact that on every solid, the top and bottom faces have the same area, as do the left and right faces and the front and back faces. Thus, the total surface area is the sum of areas that occur in pairs, which will always produce an even number. Some students may be able to generalize a formula for surface area, such as

surface area = (2 x length x width) + (2 x width x height) + (2 x length x height)

To design nets for their solids, students can think about how they might wrap their structures in wrapping paper. If they imagine placing the solid on the wrapping paper and folding up the sides of the paper to cover the four side faces and then affixing to one of the sides a piece that would cover the top, their net may look like this:

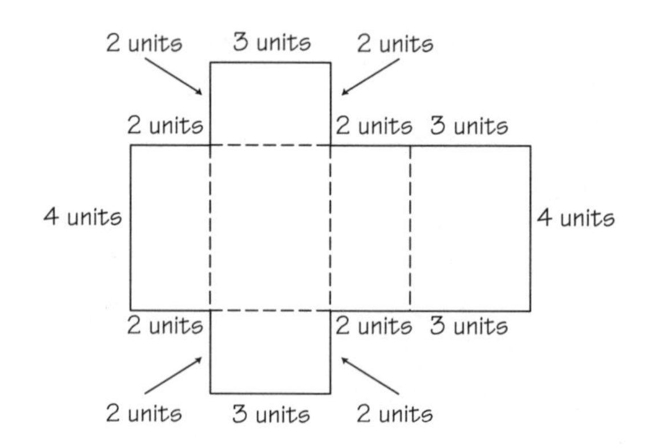

Some variations of this net pattern are shown below.

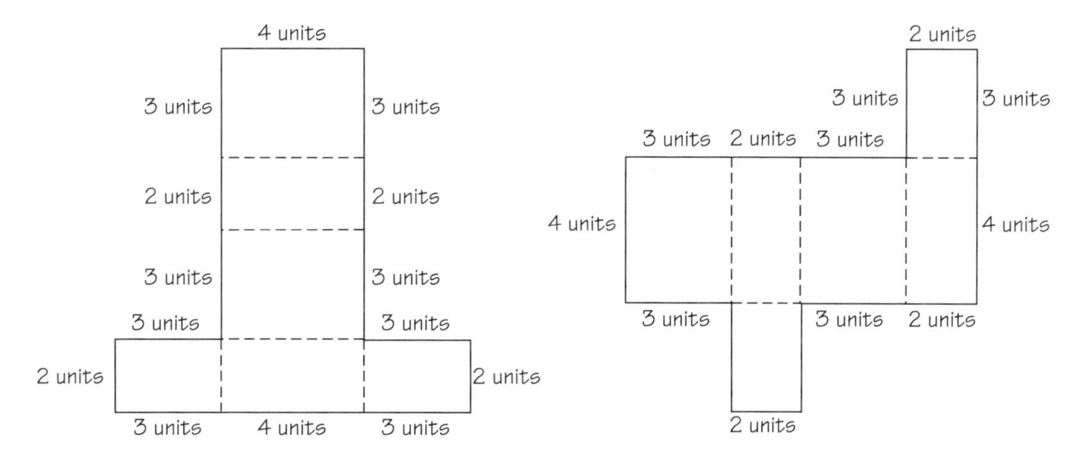

Students should find that the number of square units of wrapping paper needed to cover each of their packages is equal to the surface area of each solid. They should also recognize that the measures of the lengths of the sides of their nets are directly related to the dimensions of their solids. The class chart should reinforce that the more compact packages require less wrapping paper than the more elongated ones.

In Part 2, students extend their work to nonrectangular solids. Many will find that construction of these nets is a bit more involved than the constructions they did for their rectangular solids. Students may describe different ways of going about the process. Some may cut out pieces of wrapping paper that would cover each of the surfaces of their packages, and then arrange them on a flat surface in such a way that when adjoined and folded, they would cover the package. They may then draw a one-piece pattern in the shape of their arrangement. Other students may imagine "unfolding" their solid, as if it were hollow and made from paper. Still others may use a technique similar to that described for rectangular solids, but allowing for more than four strips of paper to wrap up and around their package. An example of a package and a net is shown here.

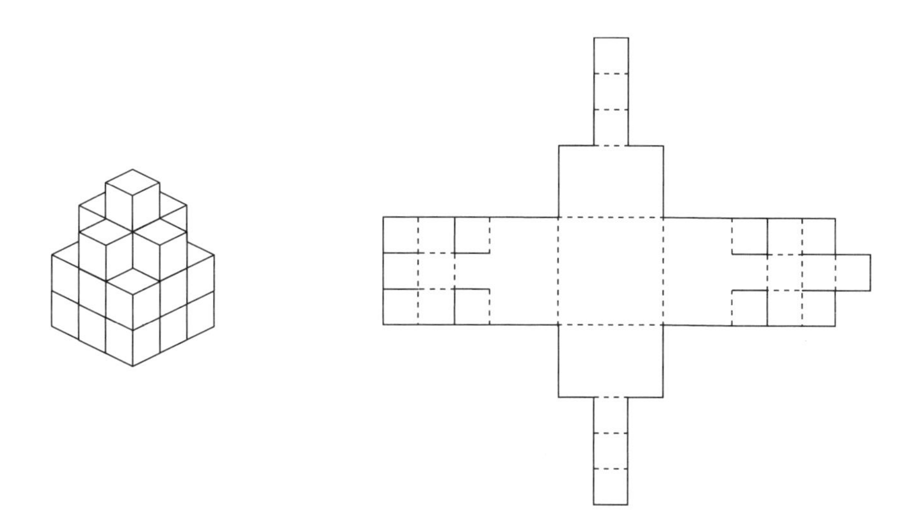

Some students may design a package in which the sides do not all form right angles. For example, the configuration of candy boxes shown in Figure A might be enclosed within the package shown in Figure B. In a case like this, students may need to use the Pythagorean theorem to calculate volume and surface area to design a net.

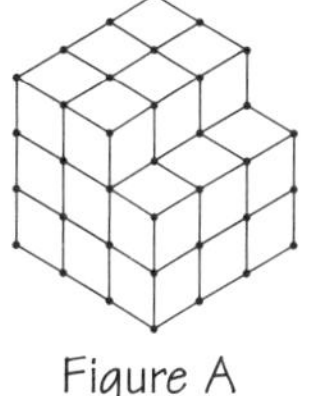

Figure A

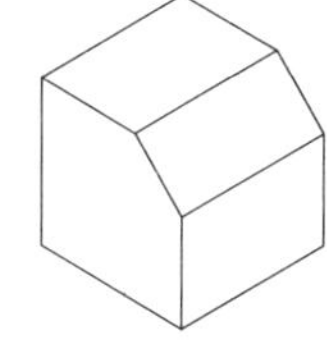

Figure B

Investigating Areas of Polygons

The lessons in this cluster focus on polygons and their areas. In particular, students explore squares and triangles on their Geoboards, and use Tangrams to investigate rectangles, nonrectangular parallelograms, and trapezoids. The first two lessons can be used as either introductory or reinforcement activities. The third lesson best serves as an introductory lesson, as students are led to a discovery of the relationships among the formulas for finding areas of the different quadrilaterals.

1. The Squarea Challenge *(Investigating squares and their areas)*

In their search to find all the different-sized squares that can be made on a Geoboard, students devise strategies for finding area and reinforce their understanding of what makes a quadrilateral a square. In *On Their Own* Part 2, students take their work one step further, as they explore ways to determine the lengths of the sides of each of their squares. Their work and the subsequent discussion can be used to introduce (or reinforce) the Pythagorean theorem and the concept of square roots.

Where's the Mathematics? (page 65) provides examples of different ways that students with varying amounts of prior knowledge may approach the activity. Teachers should encourage the sharing of different strategies during the discussion, emphasizing the value of knowing a variety of approaches to solving this and other problems.

2. Glass Triangles *(Investigating triangles and their areas)*

Through their investigation of a problem involving stained-glass windows, students investigate triangles and their areas. Students need not be familiar with the formula for calculating area of a triangle, as the nature of the Geoboard provides for other methods that can be used to find area (see *Where's the Mathematics?*, page 69).

On Their Own Part 2 presents a challenge problem in which students use their findings from the first part of the activity to create stained-glass windows that meet certain conditions. Students are encouraged to search for several solutions to the problem and to be creative in their designs.

3. Colorful Kites *(Investigating parallelograms and trapezoids and their areas)*

In this activity, students explore the relationships that exist among related quadrilaterals and their areas. In *On Their Own* Part 1, students examine rectangles and related nonrectangular parallelograms. In Part 2, a similar investigation examines trapezoids and related parallelograms. Attention is focused on how the shapes are related, and how this impacts the formulas that can be used to calculate their areas.

The lesson serves as an introduction to the formulas used to find areas of quadrilaterals. Although it requires no previous knowledge of area formulas, students with some prior exposure to these concepts can also benefit from the exploration, as it helps to clarify *why* the formulas are related.

THE SQUAREA CHALLENGE

- Area
- Square roots
- Spatial reasoning

Getting Ready

What You'll Need

Geoboards, 1 per student

Rubber bands

Geodot paper, page 123

Overhead Geoboard and/or geodot paper transparency (optional)

Calculators (optional)

Activity Master, page 108

Overview

Students search to find all the different-sized squares that can be made on a Geoboard. They then investigate ways to determine the lengths of the sides of their squares. In this activity, students have the opportunity to:

- develop strategies for finding areas of squares
- learn about square roots
- apply the Pythagorean theorem
- use logical reasoning to solve a problem

Other *Super Source* activities that explore these and related concepts are:

Glass Triangles, page 67

Colorful Kites, page 72

The Activity

On Their Own (Part 1)

Take the Squarea Challenge: How many different-sized squares can be made on a Geoboard?

- Work with a partner. Make as many different-sized squares as you can on your Geoboard. Each vertex must be a peg on the board.
- Find the area of each of your squares. Let the area of the smallest possible square be 1 square unit.
- Record each square on geodot paper and label its area.
- Be ready to explain how you know you have found all possible different-sized squares that can be made.

Thinking and Sharing

Invite pairs to show one of their squares and explain how they found its area. (You may want to have students use overhead materials for this.) Have them post their squares and label the areas. Be sure that each possible area is represented on the board.

Use prompts like these to promote class discussion:

- How did you go about searching for squares with different areas?
- How did you find the areas of your squares?
- Did you use more than one method for finding area? If so, describe the methods you used.
- How did you decide that you had found all the different-sized squares? How could you convince someone else that no others exist?

On Their Own (Part 2)

What if... ***you wanted to find the lengths of the sides of each of your squares? How might you do this?***

- Working with your partner, find the lengths of the sides of each of your squares. Let the unit of measure be the horizontal distance between two consecutive pegs in the square.
- Label the side lengths on your recordings.
- Be ready to explain the method(s) you used to find the lengths of the sides of your squares.

Thinking and Sharing

Refer to the posted squares and invite students to explain how they found the lengths of the sides of each square. Encourage students who used different methods to tell about the methods they used (for example, the Pythagorean theorem, square roots, estimation based on the formula for area of a square, and so on).

Use prompts like these to promote class discussion:

- How did you find the lengths of the sides of your squares?
- Did you use more than one method? If so, describe the methods you used.
- Are the measurements you found approximations or exact values? Explain.
- How is the length of the side of a square related to the area of the square?

Suppose your Geoboard had an extra row and column of pegs. What would be the area of the largest square you could make on your Geoboard? What would be the area of the second-largest square you could make? What would be the lengths of the sides of these squares? Use diagrams to help support your explanations.

Teacher Talk

Where's the Mathematics?

Students should find that there are eight different-sized squares that can be made on a Geoboard. The squares with areas of 1, 4, 9, and 16 square units will probably be the easiest for them to find. The squares with areas of 2, 5, 8, and 10 square units may be less obvious, as their sides are not parallel to the edges of the Geoboard.

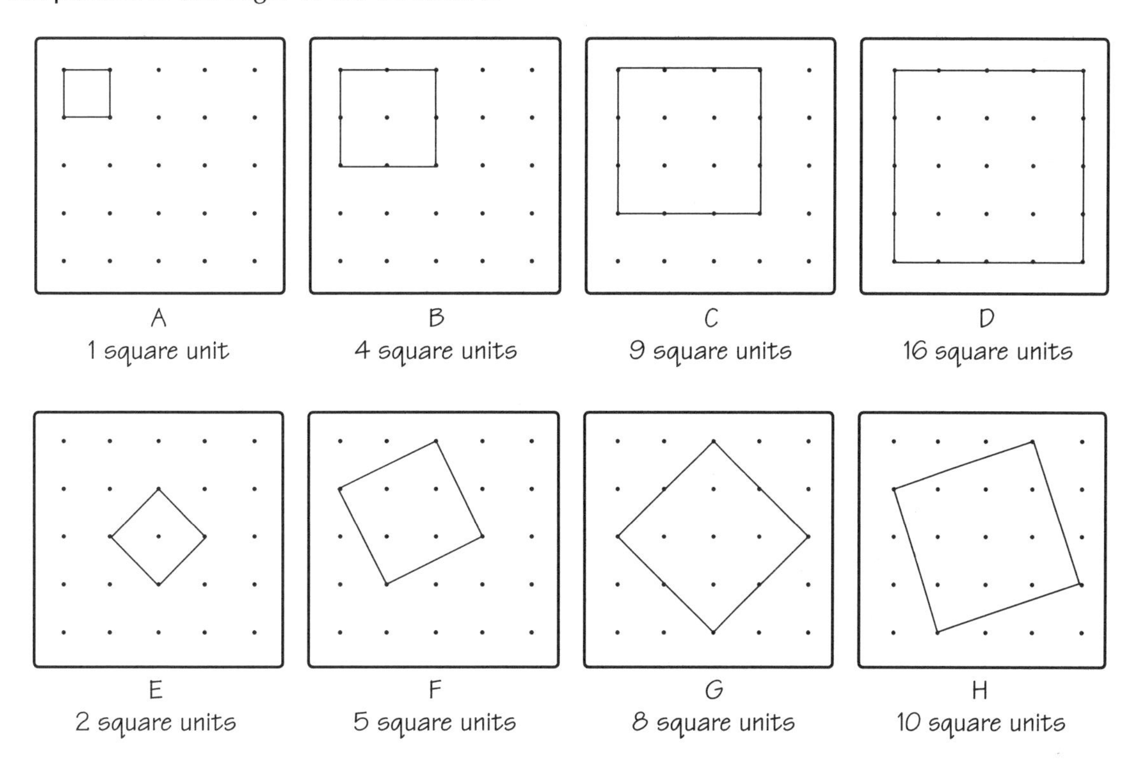

Students may use a variety of methods to find the areas of their squares. For squares A through D shown above, some students may simply count the number of 1-by-1 unit squares contained in their squares, while others may use the formula for finding area of a square: Area = side x side, or $(\text{side})^2$. These methods are not easily applied to squares E through H however. To find the areas of these squares, some students may divide their squares into unit squares and parts of unit squares, and add these areas together. For example, square G can be divided into 4 unit squares, each with an area of 1 square unit, and 8 triangles, each with an area of $\frac{1}{2}$ square unit, resulting in a total of 8 square units.

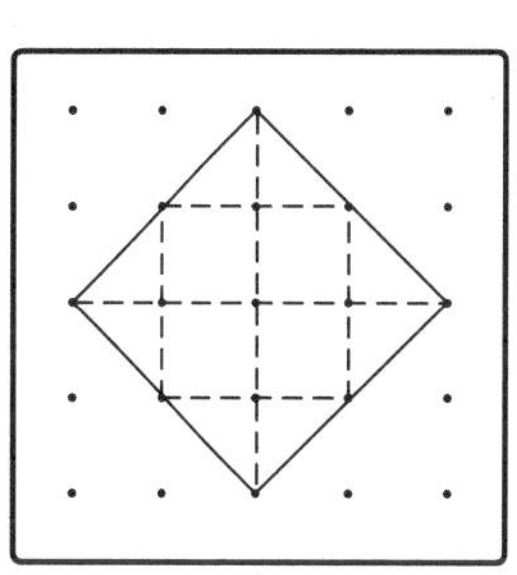

Area = $4(1) + 8(\frac{1}{2}) = 8$ square units

Some students may find the area of a "tilted" square by enclosing it in a larger square whose sides are parallel to the edges of the Geoboard. They can then find the area of the larger square, and subtract the areas of the triangular regions that lie between the two squares to find the area of the tilted square. In this example, the area of the large enclosing square is 16 square units. The area of each

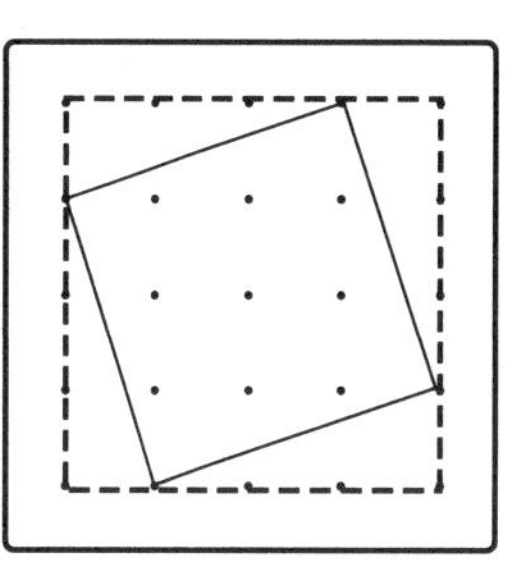

Area = $16 - 1\frac{1}{2} - 1\frac{1}{2} - 1\frac{1}{2} - 1\frac{1}{2} = 10$ square units

of the triangular regions (which can be found in a number of different ways) is $1\frac{1}{2}$ square units. Thus, the area of the original square is 10 square units.

Students may describe other methods for finding area. Some may be combinations of the techniques already described, while others may be totally different approaches.

There are also a variety of techniques that students may use to find the lengths of the sides of their squares in Part 2. For squares A through D, students can simply count the number of units. The lengths of the sides of these squares are, respectively, 1, 2, 3, and 4 units. For squares E through H, the task is a bit more challenging.

Some students may visualize the side of a "tilted" square as the hypotenuse of a right triangle with legs parallel to the sides of the Geoboard. They can then use the Pythagorean theorem to find the desired length.

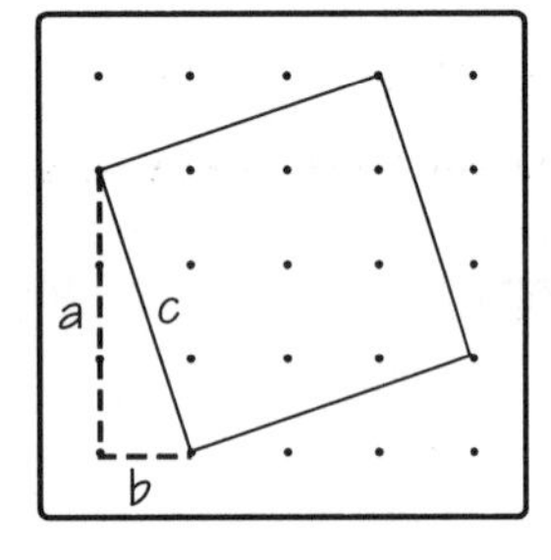

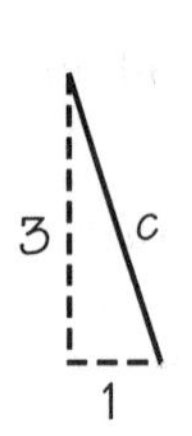

$$a^2 + b^2 = c^2$$
$$3^2 + 1^2 = c^2$$
$$9 + 1 = c^2$$
$$10 = c^2$$
$$\sqrt{10} = c$$

The length of the side is $\sqrt{10}$ units which is approximately 3.16 units

Others may approximate the length of a side by using the fact that the square of the side length must be the area of the square. They may reason, for example, that since square E has an area of 2 square units, the length of one of its sides must be greater than 1 and less than 2 (since the squares of these numbers are 1 and 4). Using trial-and-error on successive approximations, students may determine that the length of the side is about 1.41 units (an approximation of $\sqrt{2}$).

Students who are familiar with square roots may realize that since the area of a square is equal to the square of a side, the length of the side must be equal to the square root of the area. These students may use calculators to calculate approximations of the side lengths, or may represent the lengths using the square root symbol, $\sqrt{\ }$. Measurements written with the square root symbol are representations of exact values, while their decimal approximations are not.

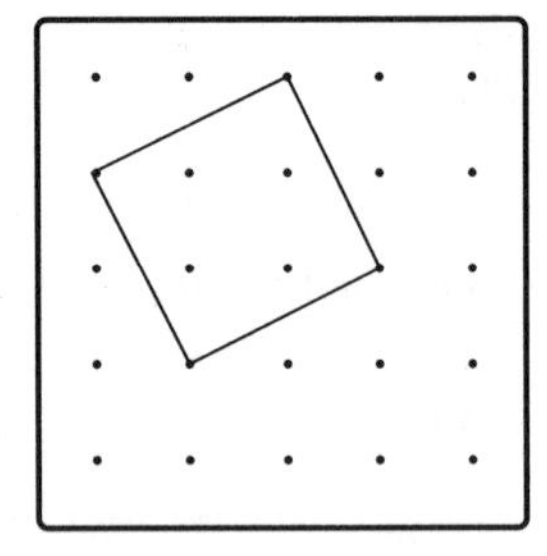

Area of square = 5 square units
Length of side = $\sqrt{5}$ units $\approx$ 2.24 units

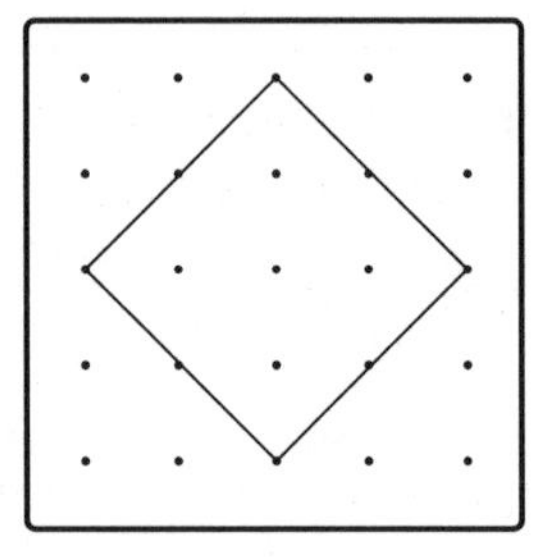

Area of square = 8 square units
Length of side = $\sqrt{8}$ units $\approx$ 2.83 units

GLASS TRIANGLES

- Area
- Congruence
- Spatial visualization

Getting Ready

What You'll Need

Geoboards, 1 per student

Rubber bands

Scissors

Geodot paper, pages 119 and 123

Activity Master, page 109

Overview

Students search to find all the possible areas of triangles that can be made on a Geoboard. They then investigate combinations of different triangles that can be used to completely cover the Geoboard. In this activity, students have the opportunity to:

- develop strategies for finding the areas of triangles
- learn that triangles with the same area are not necessarily congruent
- strengthen spatial reasoning skills

Other *Super Source* activities that explore these and related concepts are:

The Squarea Challenge, page 63

Colorful Kites, page 72

The Activity

On Their Own (Part 1)

Ernie designs stained glass. He wants to create a stained-glass window using only triangular pieces. If he uses a Geoboard as a template, how many different triangles can he make?

- Work with a group. Each of you should make a Geoboard triangle that has a different area. Use only one rubber band to make each triangle. Record your triangles on geodot paper and label the areas.
- Continue to make and record triangles until you have at least one for each of the possible areas a Geoboard triangle can have.
- Cut apart and organize your recordings.
- Be ready to explain how you conducted your search and organized your work.

Thinking and Sharing

Invite students to share their triangles and post recordings until there are several examples posted for each possible area. Discuss the methods students used to find all the possible areas and the different triangles.

Use prompts like these to promote class discussion:

- How did you go about searching for triangles with different areas?
- How did you find the areas of your triangles? Did you use more than one method? If so, describe the methods you used.
- How did you organize your work?
- How did you know you found every possible area?
- Did you see any patterns in the data you collected?
- What discoveries did you make about Geoboard triangles?

On Their Own (Part 2)

What if... *Ernie wants to create a square window containing at least five glass triangles, each having a different area? If he uses a Geoboard to model the window, what designs can he make?*

- Using your Geoboard as a frame and your triangles from Part 1, create a stained-glass design that completely fills the frame and uses at least five triangles, each with a different area.
- Be sure that there are no "holes" in your designs. The design must contain only triangles, attached side to side.
- If you find a triangle that was not on your list from Part 1, you may add it to the list and use it in your design. Record your design on geodot paper.
- Investigate other possible square window designs that could be made with your triangles. Record them on geodot paper. Be ready to discuss your work.

Thinking and Sharing

Invite students to share their designs. You may want to allow them to cut out the triangles in their designs from colored transparencies and display them using an overhead projector.

Use prompts like these to promote class discussion:

- How did you go about choosing your triangles? creating your designs?
- What was difficult about the activity? What was easy?
- How are your designs alike? How are they different?
- Did you find any triangles that were not on your list from Part 1? If so, why do you think you missed them in Part 1?
- What generalizations can you make about the possible square windows that Ernie could make with his stained-glass triangles?

Suppose Ernie's design must include a triangle that covers an area of 8 square units. How many possible designs can be created if the other 4 triangles must have different areas? Write a letter to Ernie detailing his design choices and explaining how you found them.

Teacher Talk

Where's the Mathematics?

In Part 1, students may be surprised to discover that it is possible to make Geoboard triangles with 16 different areas. One triangle for each area is shown below.

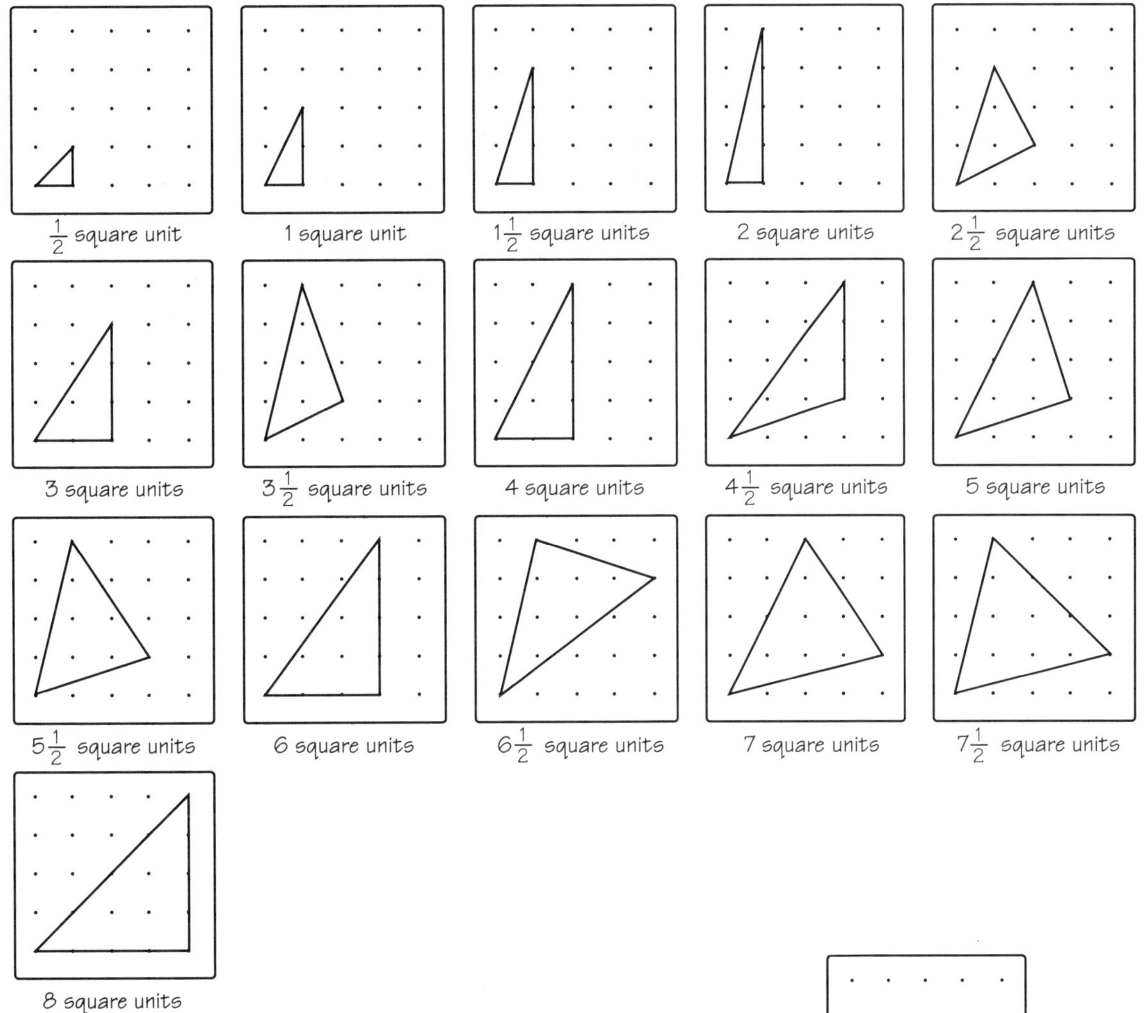

Note: Some students may point out that it is possible to make triangles with other areas by crossing the rubber band over itself, as in the example shown here.

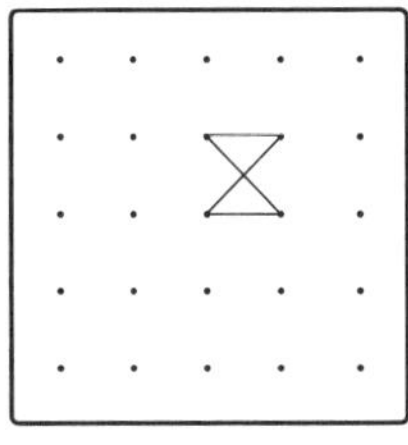

2 triangles, each with an area of $\frac{1}{4}$ square unit

As students collect and organize their triangles, they may notice that the area of the smallest possible triangle (1 x 1) is $\frac{1}{2}$ square unit and the area of the largest (4 x 4) is 8 square units. They may hypothesize that they should be able to make triangles with areas that range from $\frac{1}{2}$ square unit to 8 square units in increments of $\frac{1}{2}$ square units. This may prompt them to search for triangles having particular areas that they may have missed, and help them to determine or verify the areas of some of the triangles they were unsure about.

There are several methods students may use to find the area of their triangles. Some students may try to count the number of unit squares contained in the interior of their triangles, piecing together the partial squares. This may become difficult to do for many Geoboard triangles.

Some students may apply the formula for area of a triangle (Area = $\frac{1}{2}$ base x height). This method works well for triangles that have one or two sides parallel to the sides of the Geoboard.

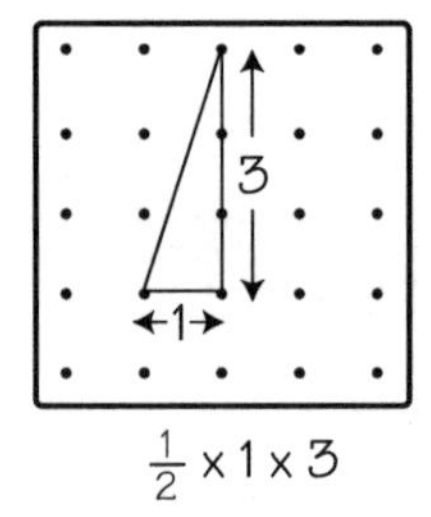

$\frac{1}{2} \times 1 \times 3$

Area = $1\frac{1}{2}$ square units

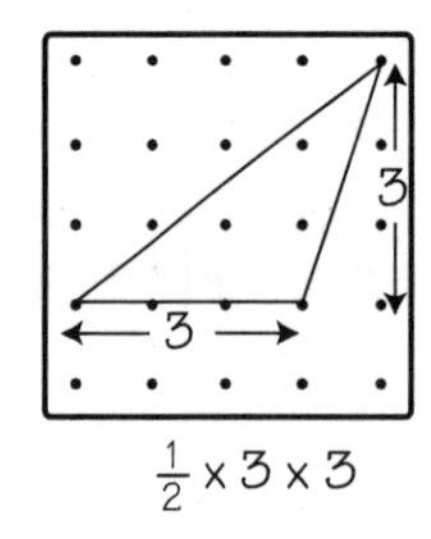

$\frac{1}{2} \times 3 \times 3$

Area = $4\frac{1}{2}$ square units

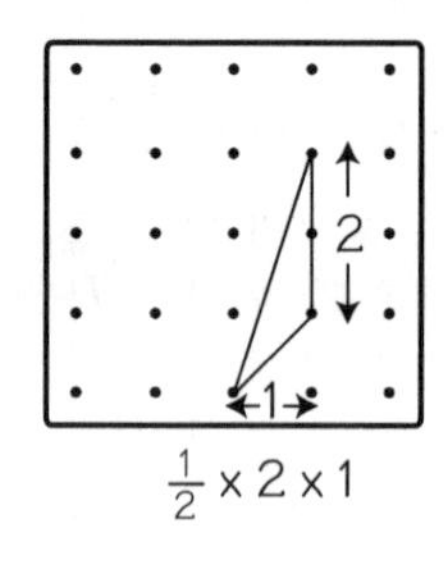

$\frac{1}{2} \times 2 \times 1$

Area = 1 square unit

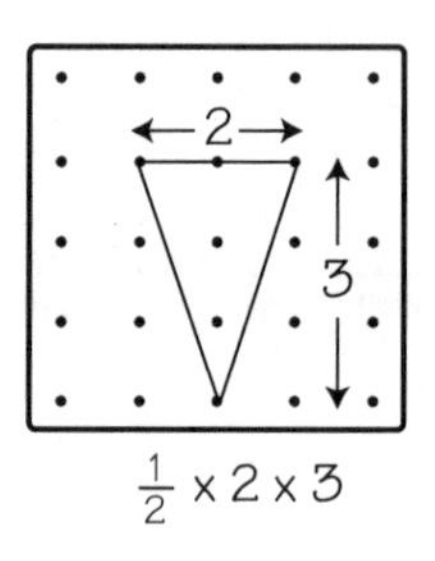

$\frac{1}{2} \times 2 \times 3$

Area = 3 square units

For other triangles, it may be easier to enclose the triangle in a rectangle whose sides are parallel to the sides of the Geoboard. Students can then find the area of their triangle by first finding the area of the rectangle, and then subtracting the areas of the right triangles surrounding the original triangle.

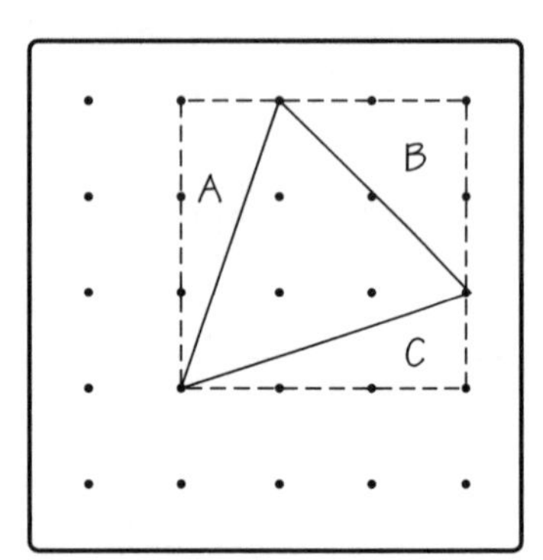

Area of rectangle = 3 x 3 = 9 square units

Area of A = $\frac{1}{2} \times 1 \times 3 = 1\frac{1}{2}$ square units

Area of B = $\frac{1}{2} \times 2 \times 2 = 2$ square units

Area of C = $\frac{1}{2} \times 3 \times 1 = 1\frac{1}{2}$ square units

Area of original triangle = $9 - (1\frac{1}{2} + 2 + 1\frac{1}{2}) = 4$ square units

Pick's theorem is another method that can be used to find the area of Geoboard shapes. The formula states that Area = $\frac{B}{2} + I - 1$, where B represents the number of pegs on the boundary of the shape, and I represents the number of pegs in the interior.

For example, the triangle shown here has 8 boundary pegs and 3 interior pegs. Applying Pick's theorem, the area of this triangle is $\frac{8}{2} + 3 - 1$, or 6, square units.

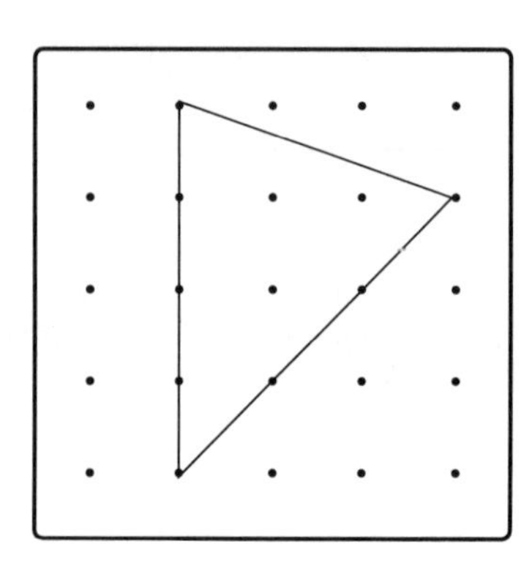

In Part 2, students must realize that the Geoboard represents a frame that contains 16 square units. Students can use this condition to decide which combinations of triangles they can use in their designs. The challenge will be finding combinations of five or more different triangles whose areas total 16 square units and whose shapes will fit together to fill the frame without overlaps or gaps. There are a variety of combinations that will work, two of which are shown below. Encourage students to search for more than one solution and to be creative in designing their windows.

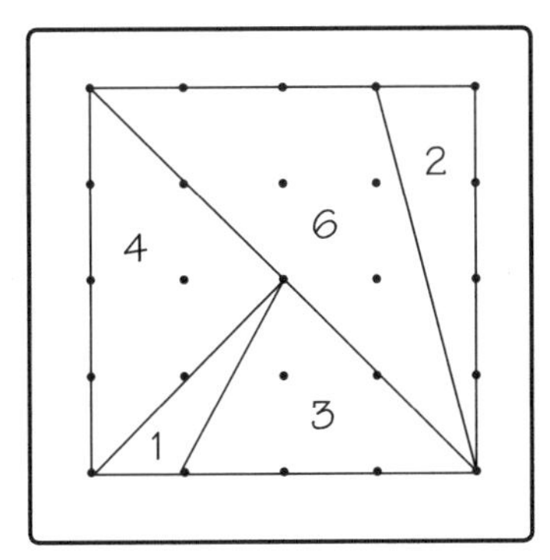

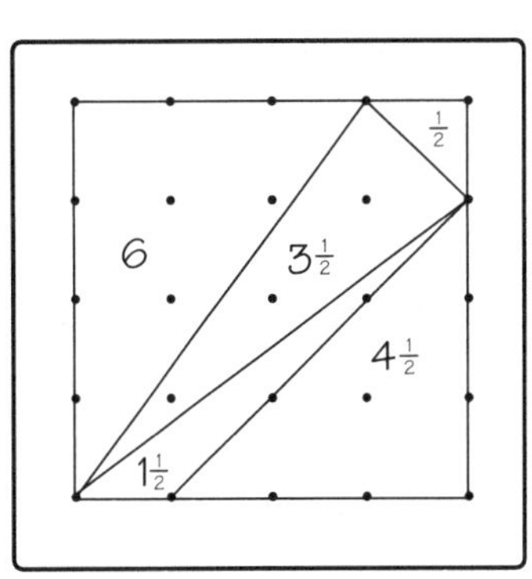

COLORFUL KITES

- Area
- Formulas
- Properties of polygons

Getting Ready

What You'll Need

Tangrams, 1 set per student

1-centimeter grid paper, page 116

Metric rulers

Activity Master, page 110

Overview

Students use Tangrams to investigate how rectangles and nonrectangular parallelograms are related, and to derive the formulas for finding the areas. In this activity, students have the opportunity to:

- see how the area of a shape is related to its dimensions
- work with metric units of measurement
- investigate the relationships that exist among rectangles, nonrectangular parallelograms, and trapezoids

Other *Super Source* activities that explore these and related concepts are:

The Squarea Challenge, page 63

Glass Triangles, page 67

The Activity

On Their Own (Part 1)

Kahlil and Kerren have decided to make colorful kites using geometric shapes cut from different-colored plastic sheets. They have a limited amount of materials and want to make the best use of them. What can you help them discover about the areas of the shapes they could use?

- Work with a partner. Each of you should make a different rectangle using between 4 and 7 Tangram pieces.
- Position your rectangle on grid paper and trace each piece. Measure and record the base and height of your rectangle (in centimeters). Then count the square centimeters covered by your rectangle to find its area, estimating where necessary.
- Now change your rectangle into a nonrectangular parallelogram using the same Tangram pieces. Trace the new shape, and find and record the base, height, and area as you did before.
- Compare the measurements of the pairs of related shapes. Be ready to discuss how your findings might influence Kahlil and Kerren in the design and construction of their kites.

Thinking and Sharing

Invite volunteers to post and label one of their rectangles and its related nonrectangular parallelogram. Have other students compare their shapes with those displayed, and resolve any discrepancies.

Use prompts like these to promote class discussion:

- What were the areas of the different rectangles you made?
- Did anyone build a rectangle that could not be made into a nonrectangular parallelogram? If so, share it.
- When you changed a rectangle into a nonrectangular parallelogram, what happened to the measure of the base? the height? the area?
- What do you notice about the posted shapes?
- How can you use your findings to write formulas for finding the area of a rectangle and the area of a nonrectangular parallelogram?
- How might your findings interest Kahlil and Kerren?

On Their Own (Part 2)

What if... *Kahlil and Kerren also want to make kites shaped like trapezoids? How will the areas of the trapezoids relate to those of the rectangles and nonrectangular parallelograms?*

- This time, you and your partner should make a different trapezoid. Remember that a trapezoid has only one pair of parallel sides.

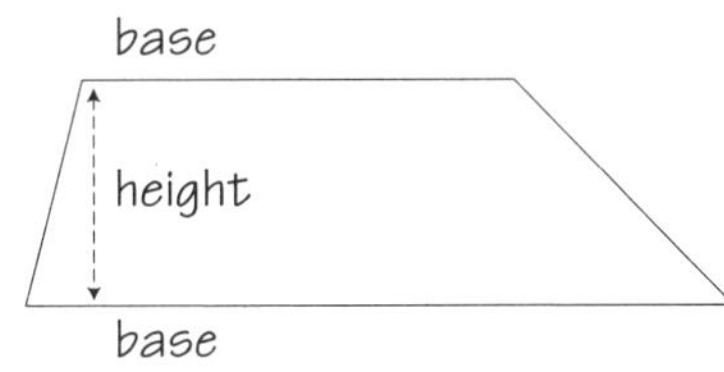

- Position your trapezoid on grid paper and trace each piece. Measure and record the length of each base, the height, and the area of your trapezoid as you did in Part 1.
- If you can, change your trapezoid into a parallelogram using the same Tangram pieces. (Note that since a rectangle is a type of parallelogram, your new shape may be a rectangle.) Trace the new shape, and find and record its base, height, and area.
- Compare the measurements of the pairs of related shapes. Use your findings to write a formula that can be used to find the area of a trapezoid when you know the lengths of its bases and its height.
- Be ready to discuss your findings.

Thinking and Sharing

As in Part 1, have students post their trapezoids and related parallelograms, labeling their dimensions and areas. Invite discussion on variations of the posted shapes.

Use prompts like these to promote class discussion:

- What do you notice about the posted shapes?

- How are the trapezoids alike? How are they different?
- Did anyone build a trapezoid that could not be made into a parallelogram? If so, share it.
- When you changed your trapezoid into a parallelogram, what happened to the height? the area? How did the length of the base of the parallelogram compare to the lengths of the two bases of the trapezoid?
- What formula did you derive for finding the area of a trapezoid? How did your work lead you to your formula?

Explain how and why the formulas for finding the area of rectangles, nonrectangular parallelograms, and trapezoids are related. Use diagrams to help illustrate your explanations.

Teacher Talk

Where's the Mathematics?

The activities in this lesson are designed to help students develop an understanding of the formulas used to find the areas of the specified quadrilaterals. Even students who are familiar with one or more of the formulas can benefit from this exploration of the relationship among the shapes, helping to clarify *why* the formulas work as they do.

As they work with the Tangrams, students may realize that the areas of related shapes are the same, because the shapes are made from the same pieces. They may also recognize that the heights are the same, although some students may need to be reminded that the height of a non-rectangular parallelogram is not the length of a side of the shape but is, instead, the perpendicular distance between bases. *They may be surprised to find that no matter how the original rectangles are transformed, the base of the new parallelogram will be the same length as that of the original rectangle. Several pairs of related shapes are shown here.

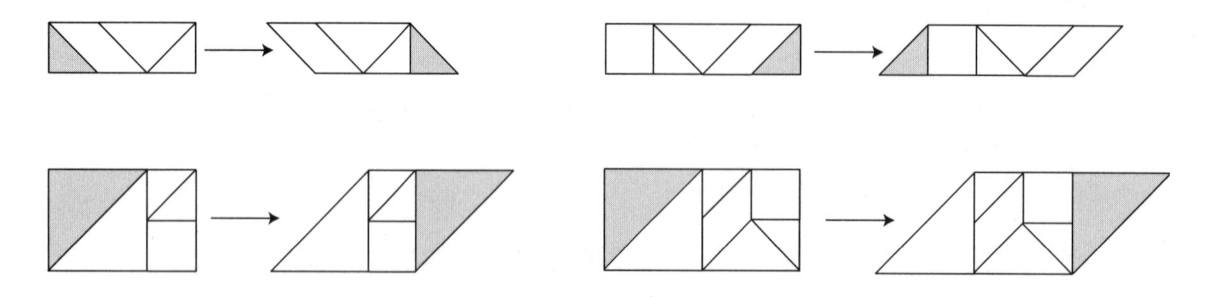

*Students may find an exception to this statement, based on the figures shown here. After the class has derived formulas based on the figures that *do* fit the statement, these new figures may spark productive discussion.

Many students will be familiar with the formula for finding the area of a rectangle: Area = base x height. (Note: Some students may use the terms *length* and *width* instead of *base* and *height.*) Once they establish that the base, height, and area of the two related shapes are the same, they can conclude that the area of the nonrectangular parallelogram can also be found by multiplying base x height. Those students that recognize that a rectangle is a parallelogram may generalize that the same formula can be used to find the area of any parallelogram.

In Part 2, students may need to investigate a number of different transformations in order to discover the relationship between the shapes. Some trapezoids are shown below.

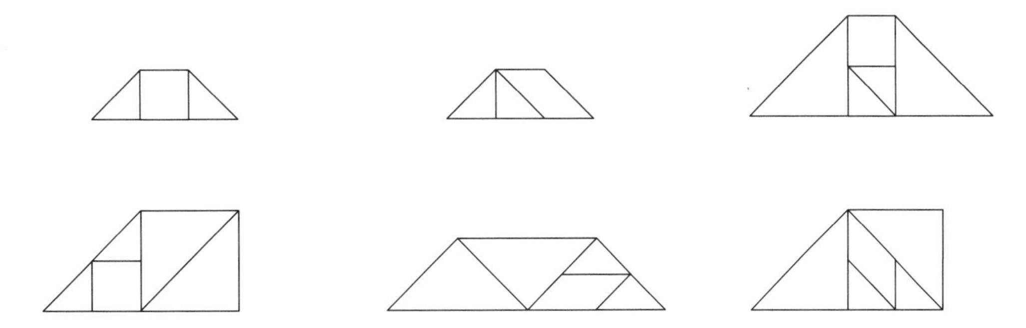

If students have not had much experience with trapezoids, they may need to review their attributes.

As they transform their trapezoids into parallelograms, students may recognize that, again, the heights and the areas of related shapes remain the same. After examining a variety of different pairs of related shapes, students may discover that the base of the related parallelogram is the average of the lengths of the two bases of the trapezoid.

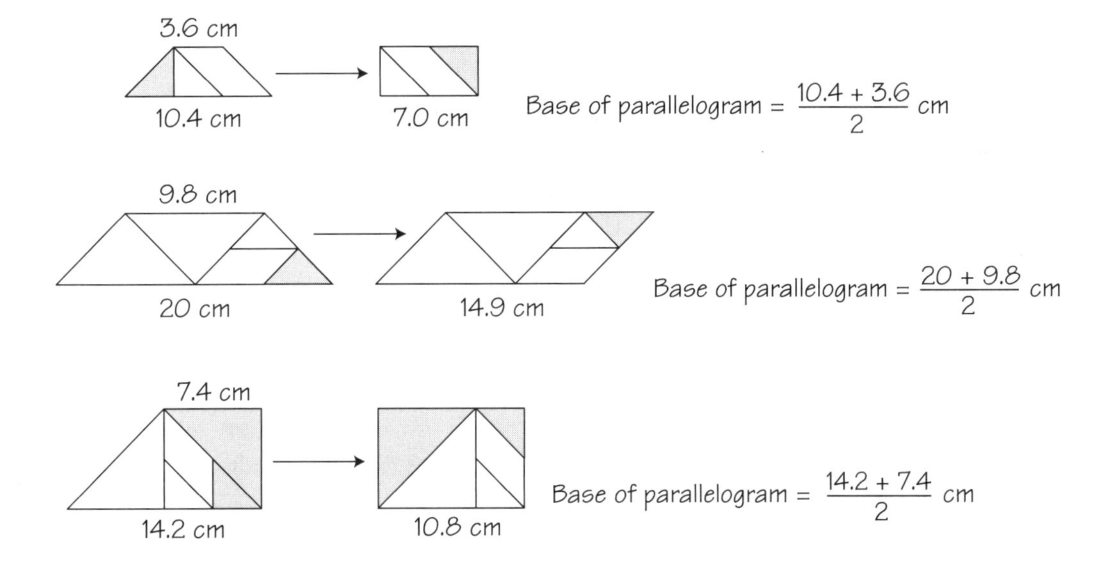

To write a formula for finding the area of a trapezoid, students can start with the formula for determining the area of a parallelogram. They can then substitute an expression for the average of the two bases of the trapezoid for the base of the parallelogram. They may express their formulas in different ways. Some of these variations appear below.

Area of a trapezoid = (average of the bases) x (height)

Area of a trapezoid = (base 1 + base 2) ÷ 2 x (height)

Area of a trapezoid = $\frac{(b1 + b2)}{2} \times h$, where $b1$ and $b2$ are the lengths of the bases, and h is the height

Investigating Angle Measure

The lessons in this cluster provide an opportunity for students to explore angles and their measures. Students investigate inscribed and central angles of circles, angles of a variety of polygons (including isosceles and equilateral triangles), and the sums of the measures of interior and exterior angles of polygons. Although the activities can be worked on in any order, it may be advisable to work on the last three lessons in the order in which they are presented.

1. Nautical Flags *(Exploring inscribed and central angles)*

Students investigate inscribed and central angles of circles motivated by their search to determine the measures of the angles of isosceles triangles they build on their circular Geoboards. They explore the relationships that exist between the angles and their intercepted arcs, and use them to determine the angle measures of their triangles. No prior knowledge of these relationships is assumed, making the lesson ideal as an introductory activity to this topic.

Before introducing the activity, teachers may want to review the definition of isosceles, and remind students that the sum of the measures of the angles of a triangle is 180°, and that a circle measures 360°. *Where's the Mathematics?,* (page 80) describes how students can use these facts in *On Their Own* Part 2 to discover that congruent chords of a circle intercept congruent arcs, and that in a triangle, the angles opposite congruent sides are congruent. The activity also provides a forum for examining the relationship between the measure of a circle (360°) and the sum of the measures of the angles of a triangle (180°), as well as for discovering properties of similar triangles.

2. Mathematical Mosaics *(Exploring angles and their measures)*

This activity provides a creative setting in which students explore the measures of angles that can be built using Pattern Blocks. They are then challenged to use their work from *On Their Own* Part 1 to design polygons that fit certain criteria. Both parts of the activity are accomplished without the use of protractors, requiring students to use their knowledge of the Pattern Block shapes to measure their angles.

This lesson is written as a Pattern Blocks activity, but a similar exploration could be done with Tangrams.

3. What's Inside? *(Exploring angles of polygons)*

In this activity, students are invited to play a game that leads them to explore the sums of the measures of the interior angles of polygons. The game rules provide elements of chance and strategy as students attempt to build polygons with the smallest possible interior angle sum.

In Part 2, students take their exploration a step further, investigating other possible sums, not only the smallest. The discussion following this part of the activity allows students to come to recognize the relationship between the number of sides of a polygon and the sum of the measures of its interior angles.

4. Interior/Exterior *(Extending angles of polygons)*

This activity provides an opportunity for students to discover the relationship between the number of sides of a polygon and the sum of the measures of its interior angles. They also investigate to determine whether such a relationship exists for the exterior angles. During the activity students are asked to make and test predictions based on their observations.

On Their Own Part 2 assumes an understanding of the distinction between convex and concave polygons. *For Their Portfolio* prompts students to consider how they might find the measure of each interior and exterior angle of a regular polygon when all that is known is the sum of the measures of the interior angles.

NAUTICAL FLAGS

- Angle measure
- Inscribed and central angles
- Isosceles triangles

Getting Ready

What You'll Need

Circular Geoboards, 1 per student

Rubber bands

Circular geodot paper, page 124

Rulers

Activity Master, page 111

Overview

Students investigate the measures of inscribed and central angles as they search for all possible isosceles triangles that can be made on a circular Geoboard. In this activity, students have the opportunity to:

- discover the relationship between the measure of inscribed and central angles and the arcs they intercept
- discover that the angles opposite the congruent sides of an isosceles triangle are congruent
- learn that in a circle, congruent chords intercept congruent arcs
- reinforce the concept of similarity

Other *Super Source* activities that explore these and related concepts are:

Mathematical Mosaics, page 82

What's Inside?, page 86

Interior/Exterior, page 91

The Activity

On Their Own (Part 1)

Jonathan makes flags that are used in marinas and on ships to indicate sailing conditions and send messages. Although the flags he makes are all different sizes, they are all isosceles triangles. If he uses a circular Geoboard to design the flags, how many different flags can he make?

- Working with your partner, create as many different-sized isosceles triangles as you can on your circular Geoboard. (Remember that isosceles triangles have at least two congruent sides.) The vertices of your triangles can be the center peg and two pegs on the circle, or they can be three pegs on the circle. (Note: Do not use the four pegs outside of the circle as your vertices.)
- Record your triangles on geodot paper. Check to make sure none of your triangles is congruent to any of your other triangles.
- Be ready to prove that each of your triangles is isosceles.

Thinking and Sharing

Invite volunteers, one at a time, to recreate one of their triangles on their Geoboard and display it on the chalk rail. Have them explain how they know their triangle is isosceles. Continue this process until all of the different triangles that students found have been displayed.

Use prompts like these to promote class discussion:

- How many isosceles triangles did you find?
- How did you know that each of your triangles was isosceles?
- Do you think that the class display shows all possible isosceles triangles that can be made on the circular Geoboard? How do you know?
- How are your triangles alike? How are they different?
- Are equilateral triangles also isosceles? Explain.
- Are isosceles triangles also equilateral? Explain.

On Their Own (Part 2)

What if... ***Jonathan needed to know the measure of the angles of the triangles so that he could program the machine that cuts the material for the flags? How could he determine these measures without the use of a protractor?***

- Using one long rubber band, make a right angle like the one shown here on your circular Geoboard. This angle is called an *inscribed angle* because its vertex lies on the circle. Record the measure of the angle.
- To find the measure of the arc intercepted by this angle, first, recall that a circle measures 360° and find the measure of the arc formed by any two consecutive pegs on the circle. Then determine the measure of the arc intercepted by your right angle.

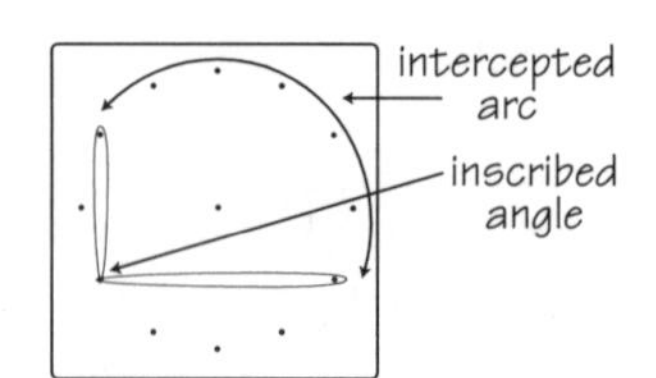

- What is the relationship between the measures of the inscribed angle and the intercepted arc? Use this relationship to determine the measures of the angles of your triangles that contain inscribed angles.
- Now make a right angle using the center peg. This angle is called a *central angle* because its vertex lies at the center of the circle. Record the measure of this angle.
- Find and record the measure of the arc intercepted by the central angle.
- What is the relationship between the measures of the central angle and the intercepted arc? Use this relationship to determine the measures of the angles of your triangles that contain central angles.
- Be ready to discuss any observations you have made about your triangles.

Thinking and Sharing

As before, invite volunteers to recreate one of their triangles on their Geoboard and display it on the chalk rail. Then have them explain how they found the angle measures. Encourage students who may have used other methods to explain their thinking. Continue until all of the different triangles have been displayed.

Use prompts like these to promote class discussion:

- What relationship did you find between the measure of an inscribed angle and the measure of its intercepted arc?
- What relationship did you find between the measure of a central angle and the measure of its intercepted arc?
- What method(s) did you use to find the measures of the angles of triangles?
- What observations did you make about your triangles?
- Do any of the triangles have the same angle measures? How would you describe these triangles?
- What generalizations can you make about isosceles triangles?

Write an explanation of how the sum of the measures of the angles of a triangle is related to the degree measure of a circle. Use words such as inscribed angle, intercepted arc, vertex, and angle measure in your explanation.

Teacher Talk

Where's the Mathematics?

The circular Geoboard provides a unique opportunity for learning about isosceles triangles and their properties. It also allows an exploration of inscribed angles, central angles, and chords of circles.

Ten different isosceles triangles can be made on a circular Geoboard. Students may need to rotate their drawings to compare their triangles and test for congruence. The triangles are shown here and on the next page.

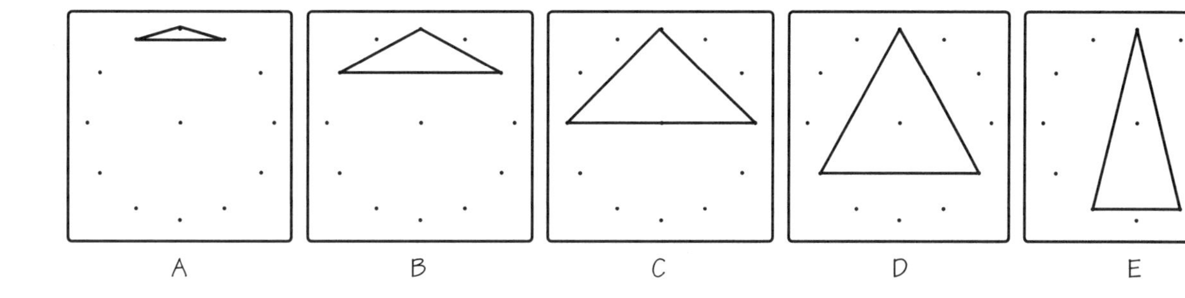

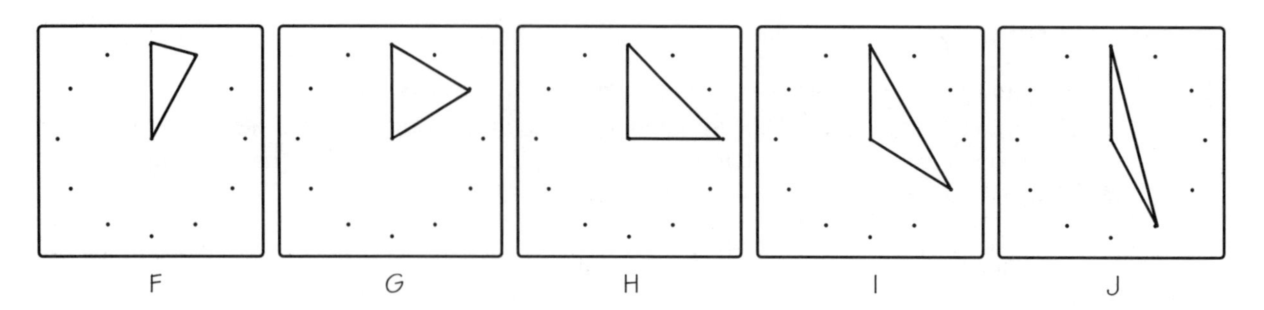

Students may have different ways of verifying that their triangles are isosceles. Some may recognize that the triangles made using the center peg have two sides that are radii of the circle, making them congruent. Some students may devise a way to measure and verify the congruence of sides that "looked like they were the same length." Others may describe a method involving counting the number of pegs between vertices to verify equivalence. For example, on the Geoboard shown, the vertices at the base of the triangle are each five pegs away from the vertex at the top of the triangle. This intuitive perception is a precursor to the idea that congruent chords of a circle intercept congruent arcs.

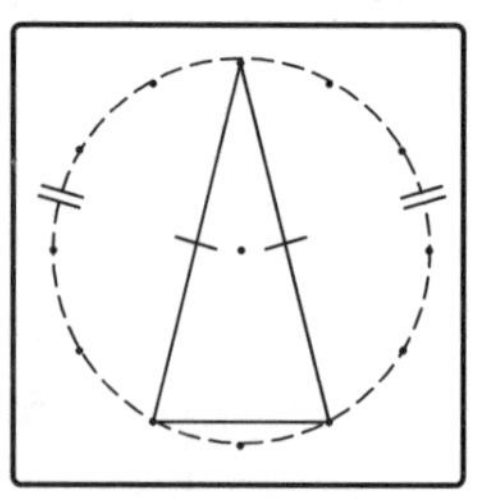
congruent chords and congruent arcs

Some students may recognize that two of the triangles are equilateral (D and G) and may question whether they are isosceles. Help them to see that they are indeed isosceles, as an isosceles triangle is defined to be a triangle with at least two congruent sides. It may help to draw the analogy that just as a square is a special type of rectangle, an equilateral triangle is a special type of isosceles triangle.

In Part 2, students should discover that an inscribed angle is half the measure of the intercepted arc, and that a central angle is equal in measure to the intercepted arc. To reach this conclusion, students must recognize that the measure of the arc formed by consecutive pegs on the circle is 30°, or 360° ÷ 12. They can then find the measures of their angles by locating the arc intercepted by each angle, finding its measure, determining the type of angle (inscribed or central), and calculating its measure accordingly. The triangles and their angle measures are shown below.

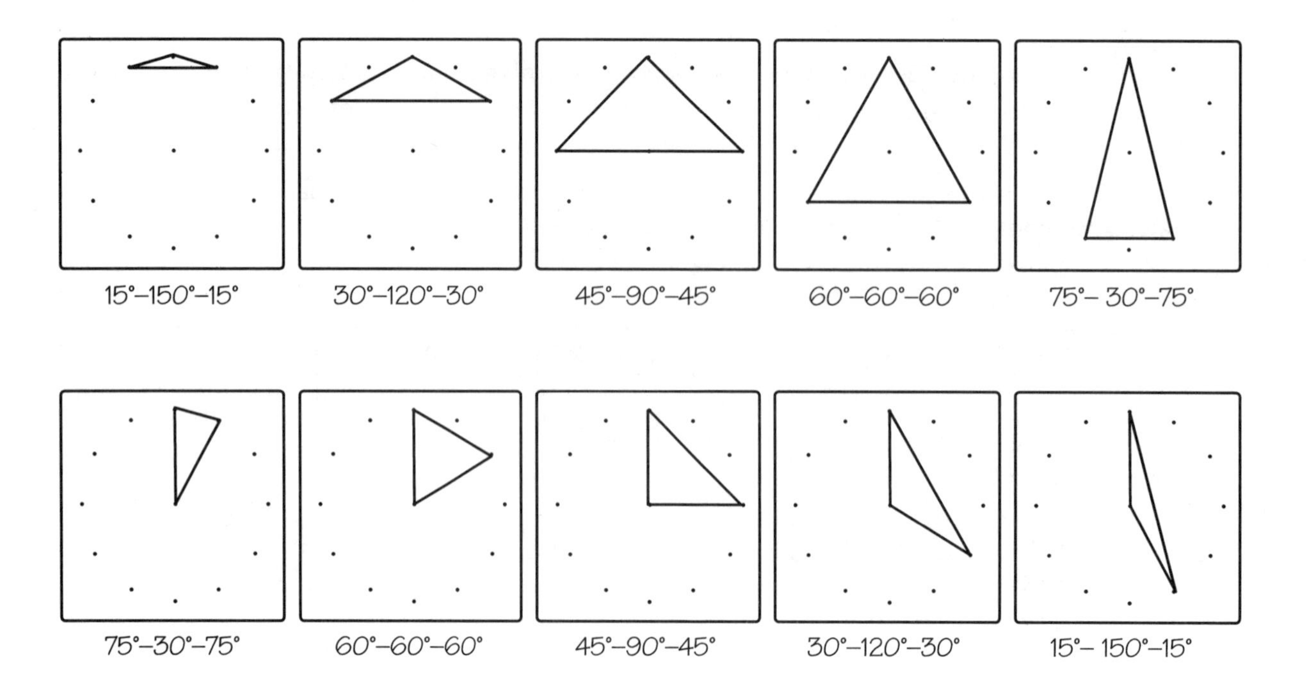

As students work, they may realize that the base angles of each triangle (the angles opposite the congruent sides) are congruent. They may then use this observation, and the fact that the sum of the measures of a triangle is 180°, to help calculate the measures of the base angles once the measure of the vertex angle has been determined. For example, if the vertex angle measures 120°, then the remaining 60° must be divided evenly between the two congruent base angles, each of them measuring 30°.

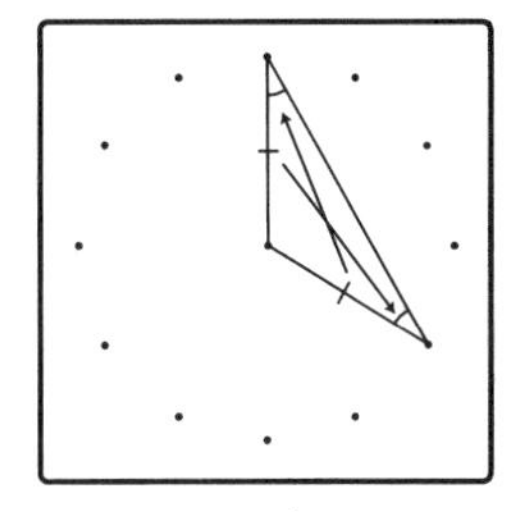

congruent angles opposite congruent sides

Some students may recognize that the three angles of an inscribed triangle, when taken together, intercept the entire circle. This observation provides a nice opportunity to point out that the sum of the angle measures (180°) is half the measure of the circle (360°), providing further validity to the relationship between inscribed angles and their intercepted arcs.

Students may be surprised to find that some of the triangles have the same angle measures. Upon closer examination, students may discover that these triangles, although different in size, have the same shape. These pairs of triangles are similar, meaning that their corresponding angles are congruent and their corresponding sides are proportional. In one pair, the triangles whose angles measure 30°-120°-30°, the triangles are not only similar, but are also congruent.

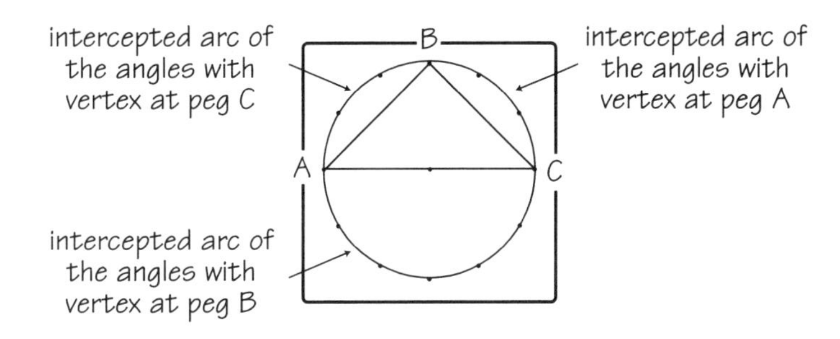

MATHEMATICAL MOSAICS

- Angle measure
- Spatial reasoning

Getting Ready

What You'll Need

Pattern Blocks, about half of 1 set per pair

Rulers

Crayons or markers

Activity Master, page 112

Overview

Students investigate the different angles that can be built using Pattern Blocks. They then use their discoveries to build polygons with different angle measures. In this activity, students have the opportunity to:

- create strategies for measuring angles
- investigate different types of angles
- search for ways to know that all possible solutions to a problem have been found
- use spatial reasoning and creativity to solve a problem

Other *Super Source* activities that explore these and related concepts are:

Nautical Flags, page 77

What's Inside?, page 86

Interior/Exterior, page 91

The Activity

On Their Own (Part 1)

Phyllis is a local artist who has decided to make a tile mosaic on the wall that borders the town playground. She decided to make the design geometric, and she will be using tiles that are in the shapes of Pattern Blocks. It would be helpful for her to know what angles are possible to make with the tiles. What can you help her discover?

- Work with a partner. Find all the different possible angles that can be made using a single Pattern Block or any combination of two or more Pattern Blocks. Use what you know about the shapes of the individual Pattern Blocks to help figure out the angle measures.
- Record each solution by tracing the blocks and labeling the angle with the measures of the individual angles (if you used more than one) and the measure of the whole angle.
- Try to build more than one arrangement of blocks for each angle measure you find.
- Be ready to explain how you know you found all possible angles.

60°
90°
150°

Thinking and Sharing

Ask students to tell what angle measures they were able to find and list them on the chalkboard. If some students were able to make angles that others could not, have them share their solutions. If there is disagreement about any of the measures, have students work together to resolve the discrepancy.

Use prompts like these to promote class discussion:

- What strategies did you use for finding new angles?
- How did you find the measures of your angles?
- What did you notice about the measures of the angles you were able to build?
- How did you decide that you had found all the possible angles?
- How did you go about building different arrangements for the same angle measure?
- Were there any angles you could build in only one way? Explain.
- What other discoveries did you make?

On Their Own (Part 2)

What if... *the mosaic is to contain polygons that have at least one angle of every possible measure? What polygons could the students use?*

- Using your work from Part 1, work with your partner to design at least three polygons for the mosaic. Your polygons may be any size and shape, but they must meet the following conditions:
    - Each polygon must contain at least one angle of every possible angle measure.
    - No polygon can contain more than three angles that have the same measure.
    - Each polygon must contain at least one tile of each of the six different Pattern Block shapes.
- Trace each of your polygons and the blocks used to make them onto white paper. Label the angle measures at each vertex.
- Be ready to tell how you went about creating the polygons for your mosaic.

Thinking and Sharing

Invite students to share their polygons and discuss how they designed them. Have them post their recordings on the chalkboard, or possibly, incorporate them into a bulletin board display.

Use prompts like these to promote class discussion:

- How did you go about designing the polygons for your mosaic?
- How did you use your work from Part 1 to help create your polygons?
- What problems did you encounter in creating the polygons? How did you solve these problems?
- What did you discover about polygons while doing this activity?

Explain how you were able to do this activity without using a protractor. Use diagrams to help clarify your explanations.

Teacher Talk

Where's the Mathematics?

Students may begin their investigation by trying to determine the measures of the angles of the different Pattern Blocks. They may use the angles of the shapes with which they are most familiar (the 90° angles of the square, the 60° angles of the equilateral triangle) to help determine the measures of the others. For example, two 60° angles fit into each angle of the hexagon, each obtuse angle of the blue rhombus, and each obtuse angle of the trapezoid, making each of these angles 120°. One 90° and one 60° angle fit into each obtuse angle of the tan rhombus, making each of these angles 150°. Other angle measures can be determined by using similar equivalence relations.

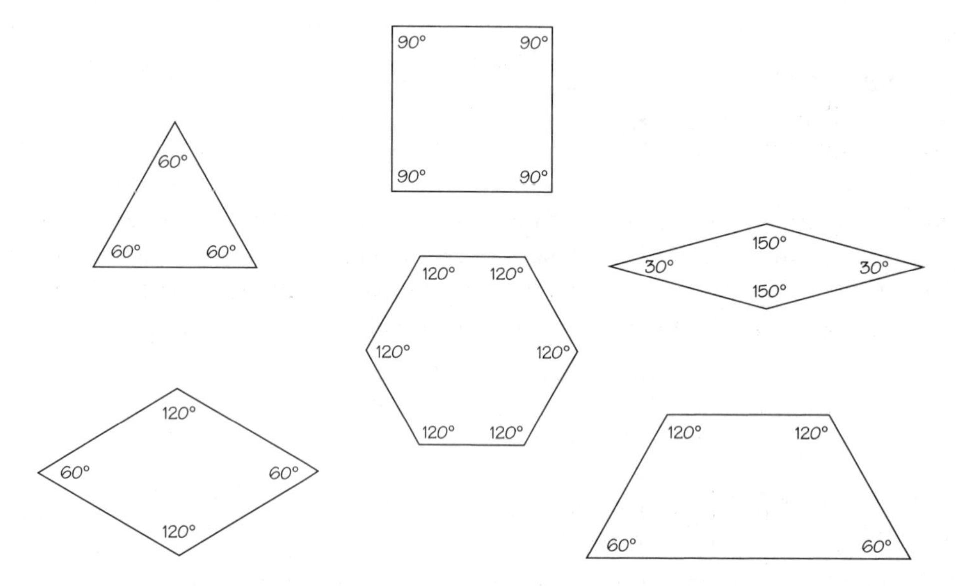

As students use combinations of blocks to build new angles, they may recognize that the measures of the angles they can build are multiples of 30°. This may help direct their search for new angles. Students should find that it is possible to make angles with measures of all the multiples of 30°, from 30° to 360°. Some examples are shown below.

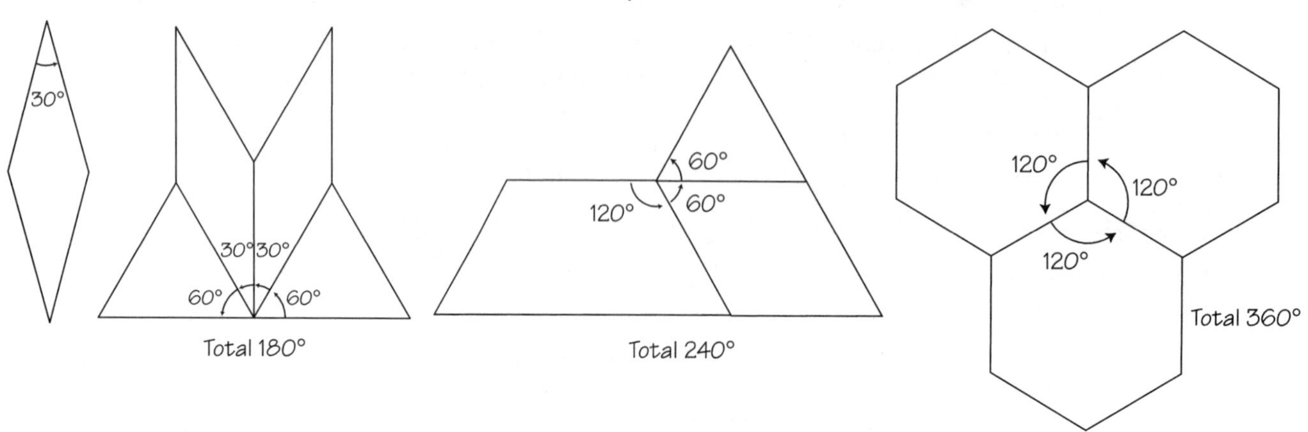

For some students, this may be their first experience with angles measuring 180° and larger. Some students may have difficulty recognizing angles that measure 180° because they may have trouble visualizing the vertex and sides of these angles. You may want to encourage these students to think of an angle as a rotation of a ray through a certain number of degrees.

Many of the angles can be represented in numerous ways. Students may use what they know about equivalence among the angles of the Pattern Blocks to make substitutions that produce angles with the same measures.

Students may offer different explanations as to how they know that they have found all the possible angles that can be made. If students went about their search in some systematic way, they may claim that they exhausted all possibilities. Students who recognized the pattern in the measures of the angles may feel convinced that they have found all possible angles once they have built angles with measures of every multiple of 30° from 30° to 360°.

The second part of the activity provides an opportunity for students to use a combination of creativity and spatial reasoning in designing polygons for their mosaics. The task is challenging, and requires students to decide on an approach, begin construction, and constantly check and modify their work. Some students may try to construct their polygons by joining together solutions from Part 1 of the activity. This may prove to be a good starting point, but adjustments may need to be made along the way. Other students may begin with a more random approach, creating a polygon with angles that appear to be different in size, and then check the angle measures and alter the shape to fit the conditions. Students may find it helpful to make a list of the possible angle measures and use it to keep track of the angle measures in their polygons.

Perhaps the most difficult requirement to fulfill is the one requiring that there be no more than three angles with the same measure in any one polygon. For an even more challenging problem, students may want to try and limit the number of angles that have the same measure to two. An intermediate problem might involve restricting the number of angle measures that appear three times to one, two, or three, and require that all other angle measures appear no more than twice. Some sample polygons and their angle measures are shown below.

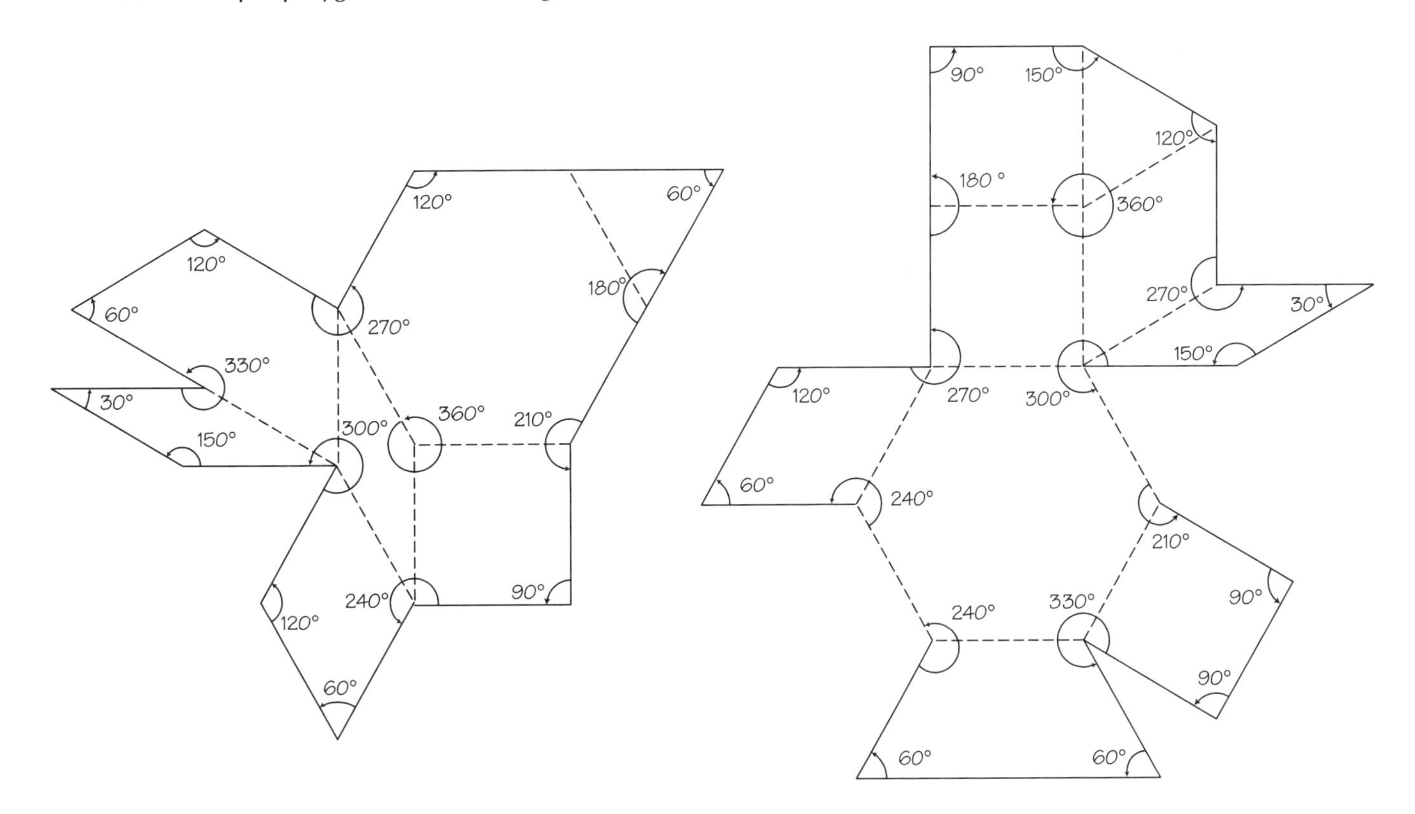

WHAT'S INSIDE?

- Angle measure
- Polygons
- Spatial visualization

Getting Ready

What You'll Need

Pattern Blocks, 1 set per group

Dice, 1 pair per group

Protractors, 1 per student

Activity Master, page 113

Overview

Students play a game in which they use different sets of Pattern Blocks to build polygons. They then investigate how the sum of the interior angles is related to the number of sides in their polygons. In this activity, students have the opportunity to:

- identify interior angles of polygons
- measure angles using a protractor
- discover properties of polygons
- develop strategic thinking skills

Other *Super Source* activities that explore these and related concepts are:

Nautical Flags, page 77

Mathematical Mosaics, page 82

Interior/Exterior, page 91

The Activity

On Their Own (Part 1)

What's Inside? is a game in which players use sets of Pattern Blocks to build polygons. The object of the game is to find the polygon whose interior angles have the smallest possible sum. Play What's Inside? following the rules below. Can you figure out the winning strategies?

- This is a game for 2 or 3 players. To begin, players take turns rolling a die 6 times, keeping track of the 6 numbers rolled. Players then assign each of their numbers to a *different* Pattern Block shape. Then players collect the assigned number of each block and place those blocks in front of them.
- For example, if a player rolls 6, 3, 2, 5, 5, 2, the player might collect:

6 yellow hexagons	3 orange squares
2 red trapezoids	5 green triangles
5 blue rhombuses	2 tan rhombuses

- Next, each player rolls the pair of dice once and adds that number of blocks to one opponent's set of blocks. The blocks do not have to be the same shape or color.
- Each player now uses all of his or her blocks to build a polygon. Blocks must be placed edge to edge, with no gaps or overlaps. The object is to try to arrange the blocks in such a way so that the sum of the interior angles of the completed polygon is as small as possible.

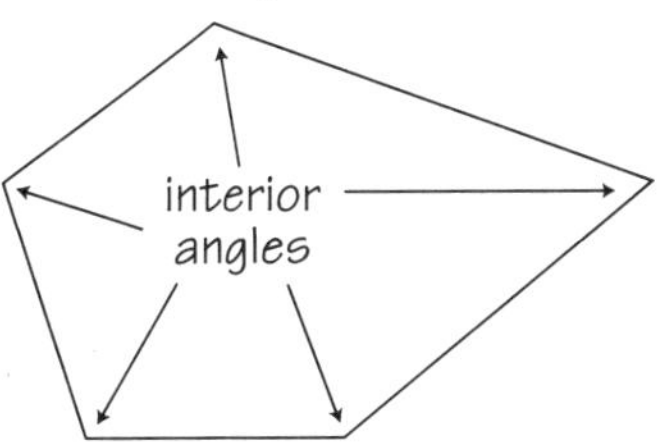

- Trace or sketch your final polygon. Label the measures of the interior angles on your drawing and record their sum.
- Play three rounds of *What's Inside?*. Add the sums from all three rounds. The player with the lowest combined total wins.

Thinking and Sharing

Invite groups to display their polygons and share their results.

Use prompts like these to promote class discussion:

- How did you decide how to assign your die rolls to the Pattern Block shapes?
- How did you decide which Pattern Blocks to add to your opponent's set of blocks?
- Do you think it is better to work with a large number of Pattern Blocks or a small number? Explain your thinking.
- What strategies did you use to build your polygons?
- What strategies did you use to calculate the measure of the interior angles? Did you find any shortcuts? If so, tell about them.
- Did you find any relationship between the size of the polygon and the sum of its interior angles? Explain.

On Their Own (Part 2)

What if... ***the rules of the game are changed a bit, and the object is now to make several polygons, each polygon having a different interior angle sum? What can you discover about the possible results of this version of the game?***

- Play *What's Inside?* with your group again, but this time, after making a polygon with the smallest possible sum, see what other sums you can obtain using the same set of blocks. Each player should try to make at least four or five polygons with different interior angle sums.
- Trace or sketch each of your polygons. Label the measures of the interior angles and record their sum.

- Compare all of the polygons recorded by the players in your group. See what you can conclude about possible interior angle sums and how they are related to the polygons that produce them.
- Be ready to discuss your findings.

Thinking and Sharing

Poll the class to determine the smallest interior angle sum found. Ask students who have recorded polygons with this sum to post their recordings. Determine the next smallest sum found, and ask students to post polygons having this sum to the right of the first set of recordings. Continue until all (or most) of the recordings have been posted.

Use prompts like these to promote class discussion:

- What do you notice about the posted shapes?
- What did you discover about possible interior angle sums?
- How does the shape of the polygon affect the sum of its interior angles?
- How does the size of the polygon affect the sum of its interior angles?
- Did you find any relationship between the number of sides of a polygon and the sum of its interior angles? If so, explain.
- How might you generalize your findings?

Suppose you are in charge of creating a new set of Pattern Block shapes. Each new shape is to have congruent sides and congruent angles. Describe how you would designate the measure of each interior angle of each block.

Teacher Talk

Where's the Mathematics?

As students play *What's Inside?*, they have an opportunity to investigate the relationship between the size and shape of a polygon and the sum of its interior angles. Students may hypothesize that the sum of the interior angles of smaller shapes will be less than that of larger shapes. Their work should prove that this is not always true. They may also experiment with convex versus concave, compact versus noncompact, or regular versus nonregular polygons to see whether these attributes affect the sum of the measures of the interior angles. Their investigation should reveal that the sum is not directly related to these factors either.

After playing several rounds of the game and analyzing the results of each round, students may come to see that the sum of the measures of the interior angles is directly related to the number of sides of the polygon. Their results should reveal that all polygons with the same number of sides have the same interior angle sum, and that the fewer the sides, the smaller the sum.

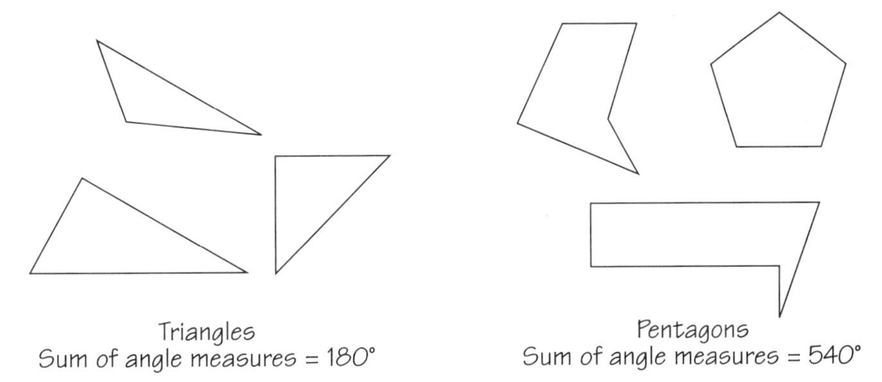

Once students discover that the best strategy for winning is to build a polygon with the fewest sides, they may investigate the best ways to do this. Students' explorations and their recognition of the spatial relationships among Pattern Blocks may lead them to use certain strategies in choosing their blocks. Some students may decide that if a set of blocks has too many squares or tan rhombuses it is difficult to minimize the number of sides. Others may find this to be the problem if the set contains too many hexagons. Students may let these observations guide them in assigning their rolls of the dice to different shapes and in considering which shapes to add to their opponents' sets.

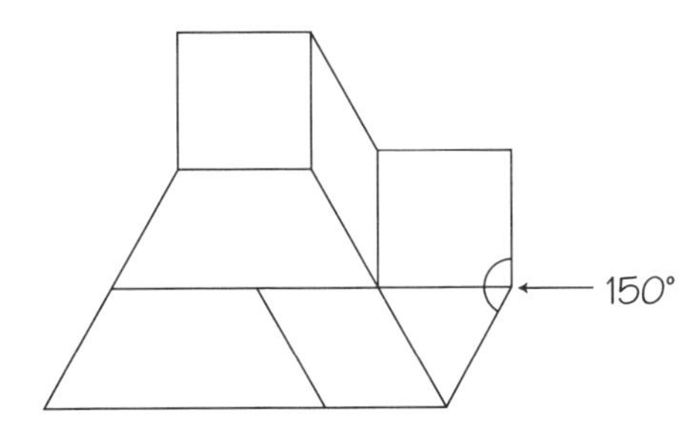

To find the measures of the interior angles of their polygons, students may use a protractor, or they may use their knowledge of the measures of the angles of the individual Pattern Blocks and add together the measures of the angles of the blocks at each vertex. For example, the angle at the vertex marked in this figure measures 150°, because it is composed of an angle of a square (90°) and an angle of an equilateral triangle (60°).

When comparing their results in Part 2, students may observe that the sums of the interior angles of their polygons are multiples of 180°. The posting of the class shapes should make the pattern even more evident: The sum of the measures of the interior angles of a triangle is 180°, of a quadrilateral is 360°, of a pentagon is 540°, and so on. Each time another side is added, the sum increases by 180°. Some students may recognize that the sum can be calculated by subtracting 2 from the number of sides in the polygon, and multiplying the result by 180°. For example, the sum of the measures of the interior angles of an octagon (8-sided polygon) would be $(8 - 2) \times 180°$, or 1080°.

Number of sides in polygon	Sum of angle measures
3	180°
4	360°
5	540°
6	720°
7	900°
8	1080°
9	1260°
10	1440°

Students may be curious about why this formula works. The formula can be easily understood by partitioning a polygon into triangles as shown below. In any polygon, the number of triangles will always be two less than the number of sides. Since the sum of the measures of the interior angles of each triangle is 180°, the sum of the measures of the interior angles of the original polygon will be the product of 180° and $(n - 2)$, where n represents the number of sides of the polygon.

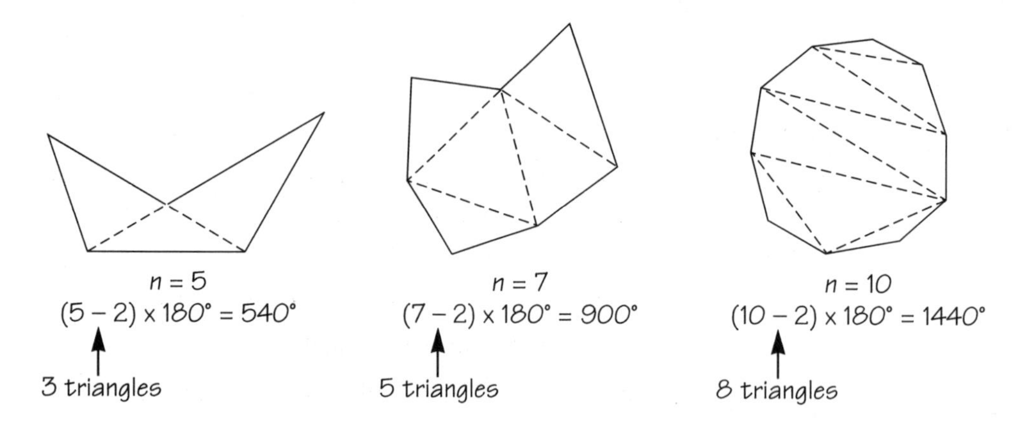

INTERIOR/ EXTERIOR

- Angles of polygons
- Convex and concave polygons
- Pattern recognition

Getting Ready

What You'll Need

Tangrams, 1 set per student

Tangram paper, page 120

Activity Master, page 114

Overview

Using Tangrams, students investigate the sums of the measures of the interior angles and exterior angles of a variety of polygons. In this activity, students have the opportunity to:

- build polygons containing various types of angles
- develop strategies for measuring angles
- distinguish between convex and concave polygons
- use patterns to make and test predictions

Other *Super Source* activities that explore these and related concepts are:

Nautical Flags, page 77

Mathematical Mosaics, page 82

What's Inside?, page 86

The Activity

On Their Own (Part 1)

Tai and Will have decided to make a book of Tangram puzzles. While drawing and measuring some of their puzzle outlines, they noticed something interesting about the angles of puzzles that have the same number of sides. Can you figure out what they discovered?

- Work with a partner. Using any combination from one set of Tangram pieces, make a shape that has four sides. Make sure your shape is different from your partner's shape. Record the outline of your shape.
- Using what you know about Tangrams, find and label the measure of each interior angle of your shape. Then record the sum of your angle measures.
- Make five more sets of shapes; each set should have a different number of sides. Record your findings as you did for the four-sided shapes.
- Organize your findings and look for patterns. Use your data to predict the sum of the measures of the interior angles of a shape with 12 sides. Check your prediction by making and measuring two different shapes that have 12 sides.
- Be ready to explain your findings and to show how you can figure out the sum of the measures of the interior angles of any polygon with a given number of sides.

Thinking and Sharing

Invite a pair of students to display one of their pairs of polygons and tell about the methods they used to find each angle measure. Have them record the number of sides of their polygons and the sum of their angle measures on the chalkboard. Repeat this for polygons having different numbers of sides. Then have students help you organize the results into a class chart, listing the polygons from those having the smallest number of sides to those having the greatest number of sides.

Use prompts like these to promote class discussion:

- How did you go about finding the measures of the angles of your shapes?
- Were some angle measures easier or harder to find than others? Which ones? Why?
- How would you describe the differences and/or similarities among the shapes?
- What patterns did you discover?
- What prediction did you make about the sum of the interior angle measures of a polygon with 12 sides? How did you formulate your prediction?

On Their Own (Part 2)

What if... ***you measured the exterior angles of your shapes? Can you find a relationship between the sum of their measures and the number of sides in the shape?***

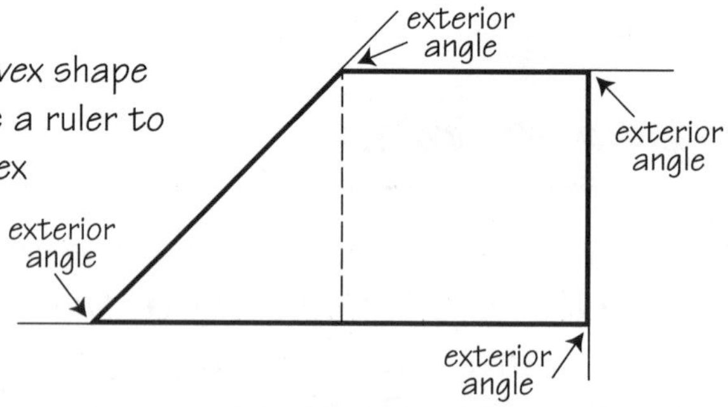

- Work with your partner. For each convex shape you recorded in the first activity, use a ruler to extend one of the sides at each vertex to form exterior angles. (If all or most of your shapes are concave, you may need to build some new shapes that are convex.)

exterior angle

exterior angle

exterior angle

exterior angle

- Determine and record the measure of each exterior angle of your shapes. Also record the number of sides of each shape and the sum of its exterior angle measures. Count only one exterior angle at each vertex.
- Organize your findings and look for patterns. Use your data to predict the sum of the exterior angles of a shape with 12 sides. Check your prediction as you did in Part 1.
- Be ready to discuss your findings about the sum of the measures of the exterior angles of a polygon.

Thinking and Sharing

Encourage students to share their methods for finding the measures of exterior angles. Ask them to describe any observations they made during their investigation.

Use prompts like these to promote class discussion:

- How did you go about finding the measures of the exterior angles of your convex polygons?
- Were some angle measures easier or harder to find than others? Which ones? Why?
- What relationship exists between each interior angle and its adjacent exterior angle?
- What patterns did you discover?
- What prediction did you make about the sum of the exterior angle measures of a convex polygon with 12 sides? How did you formulate your prediction?
- What generalizations did you make?

Suppose you know the sum of the interior angles of a particular *regular* polygon (a polygon that is both equilateral and equiangular). Explain how you might figure out the number of sides in the polygon and the measure of each interior and each exterior angle.

Teacher Talk

Where's the Mathematics?

To measure the interior angles of their polygons, students will need to determine the measures of the angles of the different Tangram pieces. Most students will recognize that the angles of the square and the right angle of each of the triangles measure 90°. They may also be able to establish that the acute angles of each triangle and the parallelogram measure 45°; perhaps by noting that any two, when taken together, are equivalent in size to one of the right angles. To determine the measure of the obtuse angle of the parallelogram, students may experiment to find that the angle is equivalent in size to an angle formed by joining a 90° angle and a 45° angle.

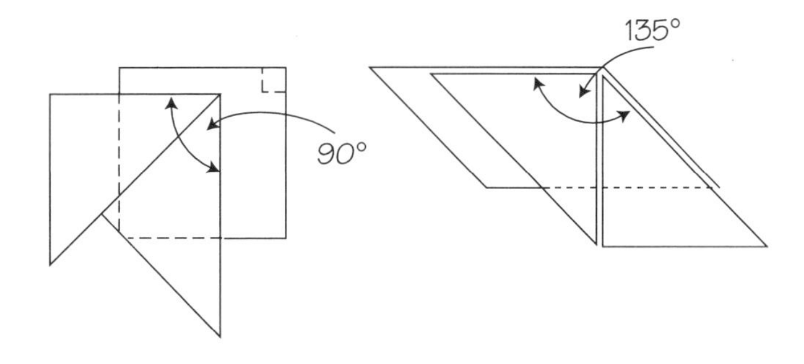

Students should find that for both of their four-sided shapes, the sum of the measures of the interior angles is 360°. Initially, they may think this to be coincidence. However, as they continue to make and measure new shapes, they may be surprised to discover that all polygons with the same number of sides have the same interior angle sums. They may test their suspicion by building very large and very small shapes with the same number of sides, and comparing their interior angle sums. They may also experiment with convex versus concave shapes. Sample shapes and their measures are shown below.

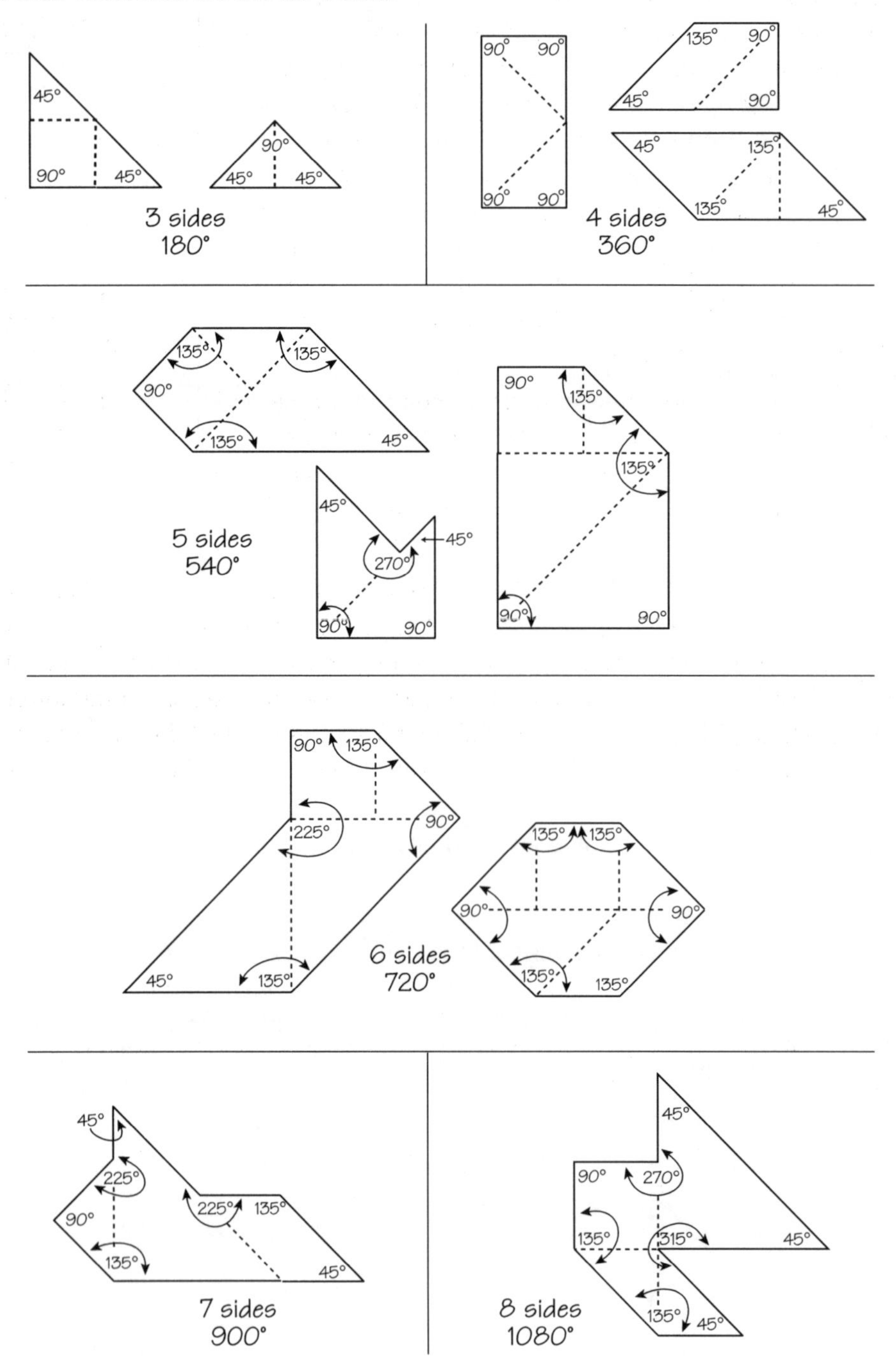

Some students may have more difficulty measuring the angles of concave polygons than of convex polygons. To measure the angles that are greater than 180°, some students may simply add the measures of the angles of the Tangrams that were used to form the angle. Other students may find the measure of the smaller angle in the exterior and subtract it from 360° (the measure of one rotation) to find the measure of the larger, interior angle.

In looking for patterns in the data, students may observe that the sums of the interior angle measures increase by 180° as the number of sides in the polygon increases by 1. By expressing the sums as repeated additions of 180°, students may discover that the number of addends of 180° is always 2 less than the number of sides in the polygon. For example, in a polygon with 7 sides, the sum of the interior angle measures is 180° + 180° + 180° + 180° + 180°, or 900°. To find the sum of the interior angle measures of a polygon with 12 sides, students can add 10 addends of 180° (or multiply 180° by 10) to obtain 1800°. Students should generalize that the sum of the interior angle measures of a polygon with n sides is $(n - 2) \times 180°$. (See page 89 of *What's Inside?* for a derivation of this formula.)

Number of sides in polygon	Sum of the interior angle measures
3	180°
4	360°
5	540°
6	720°
7	900°
8	1080°
9	1260°
10	1440°
11	1620°
12	1800°
⋮	⋮
n	$180° (n - 2)$

Students may use a variety of approaches to find the measures of the exterior angles in Part 2 of the activity. Some students may measure these angles using their Tangrams. Other students may notice that each exterior angle and its adjacent interior angle are supplementary; that is, their measures total 180°. By subtracting each known interior angle measure from 180°, the measure of the exterior angle can be found.

As students determine the measures of the exterior angles of each polygon and find their total, they may be surprised to discover that the sum is 360°, regardless of the number of sides in the original convex polygon. At every vertex of an n-sided convex polygon, the interior and exterior angles are supplementary, that is, their measures total 180°. Therefore, the sum of the interior and exterior angles of the whole polygon is $n \times 180°$. Subtract from this product the sum of the measures of the interior angles, $(n - 2) \times 180°$, and the difference, 360°, is the sum of the exterior angle measures.

$$(n \times 180°) - [(n - 2) \times 180°]$$

$$180°n - (180°n - 360°)$$

$$180°n - 180°n + 360°$$

$$360°$$

For a more intuitive explanation, suggest that students join their exterior angles at a common vertex. The angles will surround the vertex, totaling 360°.

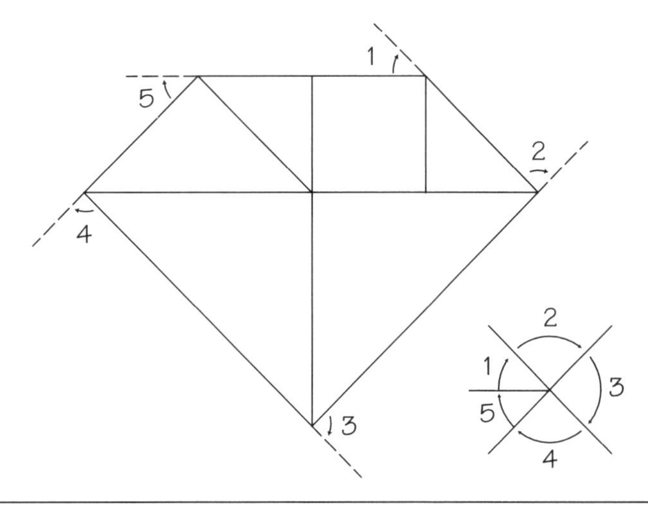

Blackline Masters

Greta's Garden

Part 1

Greta wants to build a rectangular vegetable garden. She plans to surround the garden with square patio tiles. The box of tiles she bought contains 30 tiles, each measuring 2 feet on a side. What is the largest rectangular garden plot she can surround using these tiles?

- Working with a partner, use Color Tiles to build models of possible garden border tiles. Let each Color Tile represent 1 patio tile.
- Record your models and their dimensions on grid paper. Figure out and record the perimeter and area of each of your gardens. Remember to measure the perimeter and area of the garden, not the border.
- Select the model you would use if you were Greta. Create a scale drawing of the garden you choose.
- Be ready to explain why you selected this particular model.

Part 2

What if... the garden could be any shape? Could Greta build a bigger or better garden using the same box of patio tiles?

- Consider other shapes that might be suitable for a garden.
- Build or sketch models of your ideas. You may use whatever materials you have available.
- Decide whether any of these gardens would be larger or more desirable than the rectangular garden you chose. Be ready to explain your reasoning.

Write a brief letter to Greta describing the garden you think she should build and explaining why you think it is the best choice. Include any diagrams or instructions that might be helpful.

Tiling Designs

Part 1

Marielle wants to tile the counter in her kitchen. She plans to use white tiles for most of the counter, and 26 colored tiles for a design near the center of the counter. In experimenting with different shapes, she made some interesting discoveries about the perimeters of the possible designs. What do you think she found?

- Working with a partner, make at least 10 tiling designs of different shapes, each using 26 Color Tiles. In each design, at least one complete side of each tile must touch at least one complete side of another tile.
- Record each of your designs on grid paper. Find and label the area and perimeter of each of your shapes.
- Be ready to discuss your findings about the perimeters of your designs.

Part 2

What if... the design is to be made from 28 square tiles? What could you predict about the range of possible perimeters of the different shapes that could be made?

- With your partner, discuss and record a prediction about the possible perimeters of shapes made with 28 tiles.
- Use Color Tiles to test your prediction. Modify your hypothesis if necessary.
- Be ready to prove that your hypothesis is correct.

Write a brief letter to Marielle describing the discoveries you made about the perimeters of shapes containing 26 tiles. Be sure to explain why only certain perimeters are possible.

Sandboxes

Part 1

Students from the Longview Middle School want to build a sandbox for the local playground. Using the wood they have available, how many different-sized rectangular sandboxes can they build?

- Use the following set of Cuisenaire Rods to represent the wood they have available: 2 green, 2 purple, 2 yellow, 2 dark green, and 2 black.
- Working with a partner, build models of possible rectangular sandboxes. Each sandbox must use all 10 rods for the surrounding wall. Where rods meet, they must touch along one square centimeter, not just corner to corner. Try to find sandboxes in every possible size.

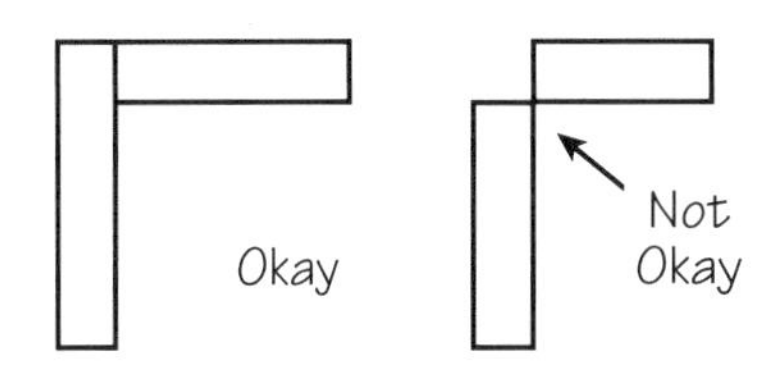

- Record your sandboxes on grid paper. Using the length of one white rod to represent 1 foot, find and record the length, width, area, and perimeter of the play area *inside* each of your sandboxes.
- Select the model you think the students should use for the sandbox. Be ready to explain why you selected this particular model.

Part 2

What if... ***the sandbox could be any shape? Could the same set of materials be used to build a sandbox that might be more desirable than a rectangular one?***

- Working with your partner, consider other shapes that may be suitable for a sandbox.
- Using the same set of rods, build and sketch models of your ideas. (Note: As in Part 1, rods much touch along one square centimeter and you must use all 10 rods for the surrounding wall.)
- Decide whether any of these sandboxes would be more desirable than the rectangular sandbox you chose in Part 1. Be ready to explain your reasoning.

Write a summary describing what would happen if there were no restrictions on the shape of your sandbox and your rods did not have to touch along one square centimeter? How would these changes effect the area and perimeter of your sandboxes? Decide which shape(s) would be preferable for a sandbox and explain why.

Perimeter Search

Part 1

During National Mathematics Week, Mr. Frangione invited students to present challenging problems to the class. Debbie posed the following question:* What are all the possible perimeters of shapes that can be made using six Pattern Blocks? *Can you solve Debbie's problem?

- Working with your partner, find all of the possible perimeters of shapes that can be made using six Pattern Blocks. You may use as many different combinations of blocks as you like.
- Use the length of the side of a green triangle as the unit of measure. Be sure to fit your blocks together so that each block shares at least one unit of length and at least one vertex with another block.
- Copy your shapes onto triangle paper and record the perimeter of each shape.
- Be ready to discuss the strategies you used to make shapes with new perimeters.

Part 2

What if... your shapes must be made using one of each of the six different Pattern Blocks? What perimeters would be possible?

- Investigate the perimeters of shapes that can be made using one of each of the six different Pattern Blocks.
- Record your shapes and their perimeters.
- Be ready to discuss and explain your findings.

Describe how you would go about finding the shapes with the smallest and greatest perimeters if you were given a set of ten Pattern Blocks.

Stadium Flip Cards

Part 1

Ruben's school is hosting a Math Olympiad. Ruben wants to make stadium flip cards that can be used by the students in the bleachers to cheer on the teams. Each student will hold 1 flip card. Can you help Ruben use a Color Tile model of the word "MATH" to determine the number of cards needed to form each flip card letter, the scale factor used, and the perimeter of each flip card letter?

MATH

- Work with a partner. Use Color Tiles to make a model of the word "MATH." Model letters must be 5 tiles wide by 5 tiles tall.
- Find the number of square units in the area of each letter, the total number of Color Tiles used, and the perimeter of each letter. Organize and record your results on a chart.
- Consider this tile arrangement to be a scale version of the students' "MATH" flip cards. If 423 students will hold cards, determine how many students are needed to hold flip cards for each of the four letters. Record these results on your chart.
- Build one or more of these new flip card letters using Color Tiles. Make sure that it is similar in design to its model. Find the perimeter(s). Add these results to your chart.
- What patterns or relationships did you find? Be ready to discuss your findings.

Part 2

What if... ***Ruben wants to create a series of flip cards that spell out "EASY AS PI"? He must first build a Color Tile model of each letter and then, based on the scale factors he wants to use, determine the area and perimeter for each of these flip card letters. (The Color Tile model letters must be 5 tiles wide by 5 tiles tall.) Can you help him with this task?***

- Using Color Tiles, build models of the letters used in the phrase "EASY AS PI."
- Find the area and perimeter of each letter. Make a chart of your findings.
- Using Color Tiles, create similar flip card letter models with a scale factor of 2.
- Find the areas and perimeters of these flip card letters and add this data to the chart under a column labeled *Scale Factor: 2.*
- Using the original Color Tile model, repeat the process to find the scale factors of 3, 4, 5, and 6, where possible. Record your findings on the chart and write about any relationships you notice in the data.
- Based on what you discovered above, consider a scale factor of *n* and *predict* the area and perimeter of the flip cards needed for the letters in "EASY AS PI."
- Be ready to discuss your findings.

Write a note to Ruben describing the change in area for a flip card letter (or any geometric shape) if the scale factor is between 0 and 1, equal to 1, or greater than 1. Include any diagrams that would be helpful.

Bon Voyage

Part 1

Tanya is planning a bon voyage party for her friend Carlos. She wants to make different-sized banners for the party using brightly colored squares of construction paper. Can you help her determine the number of paper squares needed to make each banner?

- Work with a partner. Using Color Tiles to represent the construction paper squares, build a rectangle that has a length of 4 units and width of 2 units for Tanya's first banner.
- Draw your model on the grid paper and record its dimensions and the number of construction paper tiles in its area.
- Make models for banners whose dimensions are based on changes to the original 4-by-2 banner as follows:
  (Be sure to draw the model and record the dimensions and number of tiles used for each new banner.)

  1. 4-by-6
  2. 12-by-2
  3. 12-by-6
  4. 2-by-2
  5. 4-by-1
  6. 4-by-8
  7. 2-by-1
  8. 8-by-8
  9. 8-by-6
  10. 2-by-6
- Look for relationships between the dimensions of the newly formed rectangular banners and their areas. Explain the changes in the dimensions and their affect on the area of the rectangle or the number of tiles needed to build it.
- Think about how you might generalize your findings about the area of a banner that is n times longer and m times wider. Be ready to share your findings.

Part 2

What if... ***Tanya and her friends want to buy Carlos a set of luggage as a bon voyage gift? After comparison shopping, they decide to consider how the dimensions of each piece of luggage affect its volume before making the purchase. What conclusions do you think they may have come to?***

- Work with a partner. Using combinations of Snap Cubes to represent the dimensions of the first piece of luggage, build a 5-by-1-by-4 rectangular solid.
- Draw your model on the isometric dot paper and record its dimensions and the number of cubes in its volume.
- Make models for pieces of luggage whose dimensions are based on changes to the original 5-by-1-by-4 solid as follows:
  (Be sure to draw the model and record the dimensions and number of Snap Cubes used for each new rectangular solid.)

  1. 5-by-2-by-4
  2. 5-by-1-by-8
  3. 10-by-1-by-8
  4. 10-by-2-by-8
  5. 5-by-1-by-2
  6. 10-by-3-by-2
  7. 10-by-2-by-1
  8. 4-by-1-by-8
- Look for relationships between the dimensions of the newly formed rectangular solids and their volumes. Explain the changes in the dimensions and their affect on the volume of the solid, or the number of cubes needed to build it.
- Think about how you might generalize your findings about the volume of a piece of luggage that is n times longer, m times wider, and p times taller. Be ready to share your findings.

Write a letter to Tanya explaining how the change in the volume of a piece of luggage is affected by the change(s) to the original dimensions.

Wholes and Holes

Part 1

Jamie wants to know if she can find the area of a quadrilateral built on a Geoboard, based on what she already knows about Geoboard triangles and Pick's theorem. Help her explore this question for different quadrilaterals. You will need to know Pick's theorem: Area of a Geoboard triangle = $B/2 + I - 1$, where B represents the number of border pegs and I represents the number of interior pegs.

- Work with a partner. Find the number of units in the large Geoboard square. Record this information.
- Using one rubber band, build a quadrilateral on the Geoboard in which each vertex is a peg on a different side of the Geoboard square's perimeter. Place additional rubber bands on the Geoboard to outline the surrounding triangles. Each side of the quadrilateral corresponds to one side of each triangle.
- On Geodot paper, draw a model of the Geoboard quadrilateral and the surrounding Geoboard triangles.
- Applying Pick's theorem, find the area of each surrounding Geoboard triangle and record the data for the appropriate triangle on the diagram.
- Find the area of the Geoboard quadrilateral using the data you have collected. Be ready to explain the process used.
- Apply Pick's theorem directly to the Geoboard quadrilateral to find its area and compare this to the area previously found.
- Repeat the entire process for either a Geoboard pentagon or hexagon whose vertices are pegs on the perimeter of the Geoboard square. Be ready to discuss your findings.

Part 2

What if... ***Jamie is interested in finding the area enclosed between two polygons? She calls these figures "donuts" and wonders if she can use the Geoboard and apply Pick's theorem to these new shapes. Can you help her investigate?***

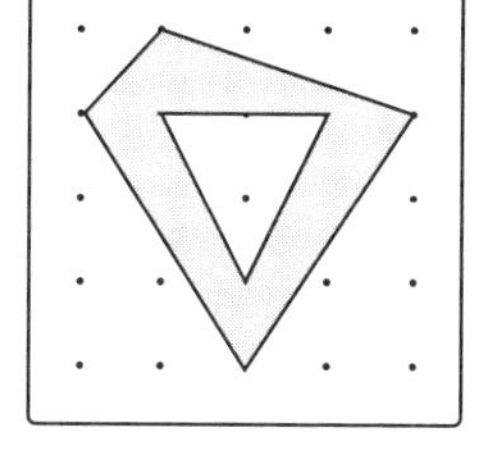

- Using the Geoboard and two rubber bands, create a donut similar to the one shown here. Be sure that the "donut hole" does not have any pegs in common with the donut's outer edge. Work so that the other groups cannot see the donut you are designing.
- Using Pick's theorem, explore ways in which the area of the donut can be determined.
- Draw a model of the Geoboard donut and record the area(s) found on the diagram.
- Exchange the value of the donut area you found with that of another group. Try to build a Geoboard donut with the given area.
- When both groups have finished, check the results against the other group's donut diagram. Discuss any difference and be ready to discuss your findings.

Write to Jamie explaining how these methods can be applied to find the areas of polygons, in general, or the area of regions located between two given shapes.

Puzzle in a Puzzle

Part 1

Sade has designed a set of Tangram challenges called "Puzzle in a Puzzle." Can you solve them?

- Working with a partner, you will need 4 sets of Tangrams to get started. Select the Tangram square from the first Tangram set; this will be your model. Then build three squares of different sizes that are similar to the model but that satisfy the following conditions:
  - the *small* square uses 3 pieces from the second Tangram set
  - the *medium* square uses 5 pieces from the third Tangram set
  - the *large* square uses all 7 pieces from the fourth Tangram set
- Assuming that the side length of the original Tangram model has a side length of 1 unit, find the side lengths of the *small*, *medium*, and *large* squares. Find and record the area of the original Tangram model and determine the areas of the *small*, *medium*, and *large* squares.
- Look for relationships between the set of squares, their side lengths, and areas.
- Now, begin a similar process by selecting the smallest Tangram triangle from the first Tangram set. Use it as the model for building three similar triangles of different sizes satisfying the following conditions:
  - the *small* triangle uses 3 pieces from the second Tangram set
  - the *medium* triangle uses 5 pieces from the third Tangram set
  - the *large* triangle uses all 7 pieces from the fourth Tangram set
- Assuming that the area of the large triangle (using all 7 pieces) has an area of 1 square unit, find and record the areas of the *model*, *small*, and *medium* triangles.
- Look for relationships between the set of triangles and their areas. Be ready to discuss your findings.

Part 2

What if... ***Sade created another "Puzzle in a Puzzle" challenge, but this time she uses Pattern Blocks instead of Tangrams? Can you solve this "Puzzle in a Puzzle"?***

- Working with a partner, use Pattern Block pieces to completely fill in the outline on the Hexagon Silhouette page. Follow the rules stated below:
  - only yellow, red, blue, and green pieces may be used
  - there must be a different number of each shape used to complete the design
- Trace the outline of each Pattern Block shape on the worksheet, labeling each piece's color.
- If the area of a blue rhombus is assumed to be 1 square unit, find the total area of the hexagon. Record your findings.
- Find the total area made up by Pattern Blocks of the same color based on the blue rhombus representing 1 square unit. Record your findings.
- Determine the fractional part or percentage of the whole area represented by the area of all Pattern Blocks of the same color.
- Be ready to share your methods and results.

Write a letter to Sade describing how you might construct a "Puzzle in a Puzzle" for her to solve. Explain or list the steps she should follow to solve your challenge.

Storage Boxes

Part 1

Kathryne takes care of her shoes by keeping them in their original shoe boxes. She wants to find one large storage box that will hold 8 of her shoe boxes. Can you help Kathryne determine the dimensions of the storage boxes that would work?

- Work with a partner. Use red Cuisenaire Rods to represent the shoe boxes.
- Arrange 8 shoe boxes so that they could fit into a rectangular prism-shaped storage box. The box should be exactly the right size to hold the 8 shoe boxes with no extra space left over. Find as many different arrangements as possible.
- Record your models on isometric dot paper. Measure and record the dimensions and volumes of each model.
- Determine the *actual* dimensions of each of Kathryne's shoe boxes if 1 centimeter in your model represents 15 centimeters for the *actual* shoe boxes. Then calculate the actual dimensions and volumes of the storage boxes that you modeled.
- Be ready to discuss your findings.

Part 2

What if... ***Kathryne decides to make her own storage box from sheets of plywood? What is the least amount of plywood she would need to create the box? What is the most?***

- Using your models from Part 1, determine the amount of plywood needed to construct each possible storage box. Record your measurements (in square centimeters) near your drawings.
- Determine which arrangement would need the least amount of plywood and which would need the most.
- Now use your observations to determine the least and greatest amounts of plywood needed to construct a storage box that would hold 12 shoe boxes.
- Try to come up with a general rule that could be used to predict what kinds of box arrangements will use the least amount of plywood.
- Be ready to explain your methods and discuss your findings.

Imagine that the storage container is to be made of cardboard instead of plywood. Using your results from Part 2, make a scale drawing of a one-piece pattern that can be cut out and folded to form the storage box needing the least amount of cardboard. Use the same scale as you used for your models. Label your pattern with the actual measurements that would be needed to construct the storage box.

Cube Sculptures

Part 1

The students in the eighth grade art class are using recyclable materials to build cubes which they will join together to make sculptures for the school courtyard. Each sculpture will be made from 16 cubes. The visible faces of each cube will be painted in different colors. How many different colors are needed?

- Work with a partner. Use Snap Cubes to design models of several different sculptures, each containing 16 cubes.
- Determine the number of colors that would be needed to paint each sculpture. Remember, each visible face must be a different color.
- Record your sculptures on isometric dot paper. Record the volume and surface area of each sculpture using the edge of one cube as the unit of measure.
- Now try to make models of sculptures that will require different numbers of colors from those you recorded. When you find one, record it as you did before.
- Continue until you think you've modeled at least one sculpture for every possible number of visible faces. Be ready to discuss your findings.

Part 2

What if... ***the students decide that each sculpture is to be enclosed in a clear plastic rectangular prism that will protect it from the weather. What size prisms will they need to construct?***

- Reconstruct one of your sculpture models. Imagine enclosing it in the smallest possible rectangular prism that could hold it.
- Determine the dimensions and volume of the prism. Record these measurements near the drawing of your sculpture.
- Reconstruct each of your other models and determine the dimensions and volumes of the prisms they would require. Record your findings.
- Compare your different models and the dimensions of their enclosing prisms. Be ready to discuss your observations.

Suppose you were building sculptures made from 20 cubes. Describe how you would construct the models having the smallest and greatest possible surface areas. Explain how you know that no other sculpture could have a smaller (or greater) surface area than the ones you described.

Wrapping Paper

Part 1

A candy company packages its candy in individual boxes, which are then wrapped in larger packages containing two dozen small boxes. If the larger packages must be rectangular solids, how many different-sized packages are possible and what patterns can be used to construct them?

- Work with a partner. Imagine that a Snap Cube represents a box that contains one piece of candy. Build three different-sized packages, each containing two dozen of the small candy boxes. Remember that your packages must be in the shape of rectangular solids.
- Find the volume and surface area of each of your packages. Use the length of an edge of one cube as the unit of measure.
- Record each of your packages and its volume and surface area on isometric dot paper.
- Design a one-piece pattern that could be folded up to form the wrapping paper for one of your packages. (You may want to use Snap-Cube grid paper as a tracing guide.) Use dashed lines to indicate where your pattern should be folded once it is cut out. There should not be any overlapping paper.
- Make a copy of your pattern. Cut it out and fold it around your solid to check that it works. If it doesn't, modify your pattern until it does.
- Label your pattern, indicating the length of each side in numbers of units. Record the total number of square units of wrapping paper needed to make your pattern.
- Repeat the process for your other packages. Organize and compare the measurements of each of your packages and be ready to discuss your findings.

Part 2

What if... ***the company decides to try packaging their candy boxes in packages that are not rectangular solids? What shapes might they consider and what nets can be used to construct them?***

- Using Snap Cubes, work with your partner to design a package that will hold two dozen candy boxes and is not a rectangular solid. Record your package on isometric dot paper.
- Determine the volume and surface area of your package.
- Design a net for your package, using dashed lines to indicate the fold lines. Then make a copy of it, cut it out, and see if it works. If it doesn't, modify it until it does.
- Label your net, indicating the length of each side in numbers of units. Also record the total number of square units of wrapping paper needed to make your net.
- Now make two different nets that could be used to wrap the same package. Again, make copies, cut them out, try them, and modify them if necessary. Label them as you did the first net. Be ready to discuss your observations.

Suppose you know the length, width, and height of a rectangular solid, but you do not have a model. Explain how you would design a net for the solid, and how you would determine how much paper would be needed to make the net.

The Squarea Challenge

Part 1

Take the Squarea Challenge: How many different-sized squares can be made on a Geoboard?

- Work with a partner. Make as many different-sized squares as you can on your Geoboard. Each vertex must be a peg on the board.
- Find the area of each of your squares. Let the area of the smallest possible square be 1 square unit.
- Record each square on geodot paper and label its area.
- Be ready to explain how you know you have found all possible different-sized squares that can be made.

Part 2

What if... ***you wanted to find the lengths of the sides of each of your squares? How might you do this?***

- Working with your partner, find the lengths of the sides of each of your squares. Let the unit of measure be the horizontal distance between two consecutive pegs in the square.
- Label the side lengths on your recordings.
- Be ready to explain the method(s) you used to find the lengths of the sides of your squares.

Suppose your Geoboard had an extra row and column of pegs. What would be the area of the largest square you could make on your Geoboard? What would be the area of the second-largest square you could make? What would be the lengths of the sides of these squares? Use diagrams to help support your explanations.

Glass Triangles

Part 1

Ernie designs stained glass. He wants to create a stained-glass window using only triangular pieces. If he uses a Geoboard as a template, how many different triangles can he make?

- Work with a group. Each of you should make a Geoboard triangle that has a different area. Use only one rubber band to make each triangle. Record your triangles on geodot paper and label the areas.
- Continue to make and record triangles until you have at least one for each of the possible areas a Geoboard triangle can have.
- Cut apart and organize your recordings.
- Be ready to explain how you conducted your search and organized your work.

Part 2

What if... ***Ernie wants to create a square window containing at least five glass triangles, each having a different area? If he uses a Geoboard to model the window, what designs can he make?***

- Using your Geoboard as a frame and your triangles from Part 1, create a stained-glass design that completely fills the frame and uses at least five triangles, each with a different area.
- Be sure that there are no "holes" in your design. The design must contain only triangles, attached side to side.
- If you find a triangle that was not on your list from Part 1, you may add it to the list and use it in your design. Record your design on geodot paper.
- Investigate other possible square window designs that could be made with your triangles. Record them on geodot paper. Be ready to discuss your work.

Suppose Ernie's design must include a triangle that covers an area of 8 square units. How many possible designs can be created if the other 4 triangles must have different areas? Write a letter to Ernie detailing his design choices and explaining how you found them.

Colorful Kites

Part 1

Kahlil and Kerren have decided to make colorful kites using geometric shapes cut from different-colored plastic sheets. They have a limited amount of materials and want to make the best use of them. What can you help them discover about the areas of the shapes they could use?

- Work with a partner. Each of you should make a different rectangle using between 4 and 7 Tangram pieces.
- Position your rectangle on grid paper and trace each piece. Measure and record the base and height of your rectangle (in centimeters). Then count the square centimeters covered by your rectangle to find its area, estimating where necessary.
- Now change your rectangle into a nonrectangular parallelogram using the same Tangram pieces. Trace the new shape, and find and record the base, height, and area as you did before.
- Compare the measurements of the pairs of related shapes. Be ready to discuss how your findings might influence Kahlil and Kerren in the design and construction of their kites.

Part 2

What if... ***Kahlil and Kerren also want to make kites shaped like trapezoids? How will the areas of the trapezoids relate to those of the rectangles and nonrectangular parallelograms?***

- This time, you and your partner should make a different trapezoid. Remember that a trapezoid has only one pair of parallel sides.

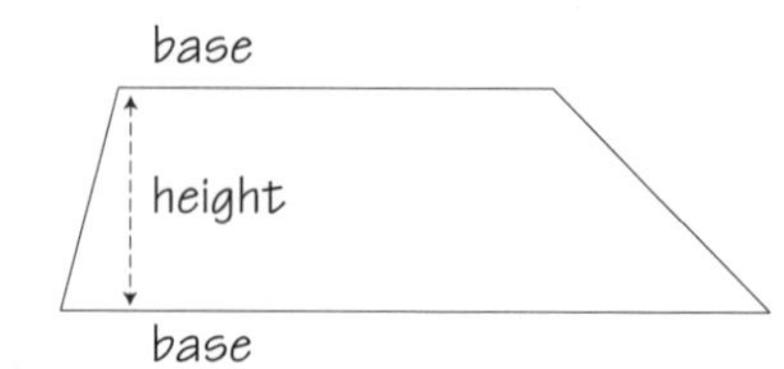

- Position your trapezoid on grid paper and trace each piece. Measure and record the length of each base, the height, and the area of your trapezoid as you did in Part 1.
- If you can, change your trapezoid into a parallelogram using the same Tangram pieces. (Note that since a rectangle is a type of parallelogram, your new shape may be a rectangle.) Trace the new shape, and find and record its base, height, and area.
- Compare the measurements of the pairs of related shapes. Use your findings to write a formula that can be used to find the area of a trapezoid when you know the lengths of its bases and its height.
- Be ready to discuss your findings.

Explain how and why the formulas for finding the area of rectangles, nonrectangular parallelograms, and trapezoids are related. Use diagrams to help illustrate your explanations.

Nautical Flags

Part 1

Jonathan makes flags that are used in marinas and on ships to indicate sailing conditions and send messages. Although the flags he makes are all different sizes, they are all isosceles triangles. If he uses a circular Geoboard to design the flags, how many different flags can he make?

- Working with your partner, create as many different-sized isosceles triangles as you can on your circular Geoboard. (Remember that isosceles triangles have at least two congruent sides.) The vertices of your triangles can be the center peg and two pegs on the circle, or they can be three pegs on the circle. (Note: Do not use the four pegs outside of the circle as your vertices.)
- Record your triangles on geodot paper. Check to make sure none of your triangles is congruent to any of your other triangles.
- Be ready to prove that each of your triangles is isosceles.

Part 2

What if... ***Jonathan needed to know the measure of the angles of the triangles so that he could program the machine that cuts the material for the flags? How could he determine these measures without the use of a protractor?***

- Using one long rubber band, make a right angle like the one shown here on your circular Geoboard. This angle is called an *inscribed angle* because its vertex lies on the circle. Record the measure of the angle.
- To find the measure of the arc intercepted by this angle, first, recall that a circle measures 360° and find the measure of the arc formed by any two consecutive pegs on the circle. Then determine the measure of the arc intercepted by your right angle.

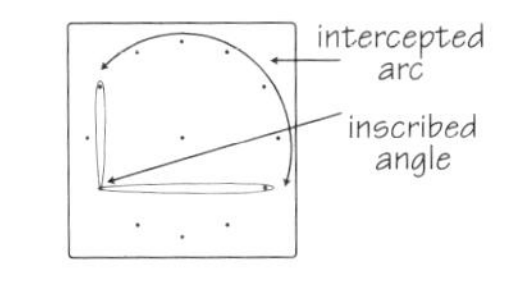

- What is the relationship between the measures of the inscribed angle and the intercepted arc? Use this relationship to determine the measures of the angles of your triangles that contain inscribed angles.
- Now make a right angle using the center peg. This angle is called a *central angle* because its vertex lies at the center of the circle. Record the measure of this angle.

intercepted arc
central angle

- Find and record the measure of the arc intercepted by the central angle.
- What is the relationship between the measures of the central angle and the intercepted arc? Use this relationship to determine the measures of the angles of your triangles that contain central angles.
- Be ready to discuss any observations you have made about your triangles.

Write an explanation of how the sum of the measures of the angles of a triangle is related to the degree measure of a circle. Use words such as inscribed angle, intercepted arc, vertex, and angle measure in your explanation.

Mathematical Mosaics

Part 1

Phyllis is a local artist who has decided to make a tile mosaic on the wall that borders the town playground. She decided to make the design geometric, and she will be using tiles that are in the shapes of Pattern Blocks. It would be helpful for her to know what angles are possible to make with the tiles. What can you help her discover?

- Work with a partner. Find all the different possible angles that can be made using a single Pattern Block or any combination of two or more Pattern Blocks. Use what you know about the shapes of the individual Pattern Blocks to help figure out the angle measures.
- Record each solution by tracing the blocks and labeling the angle with the measures of the individual angles (if you used more than one) and the measure of the whole angle.

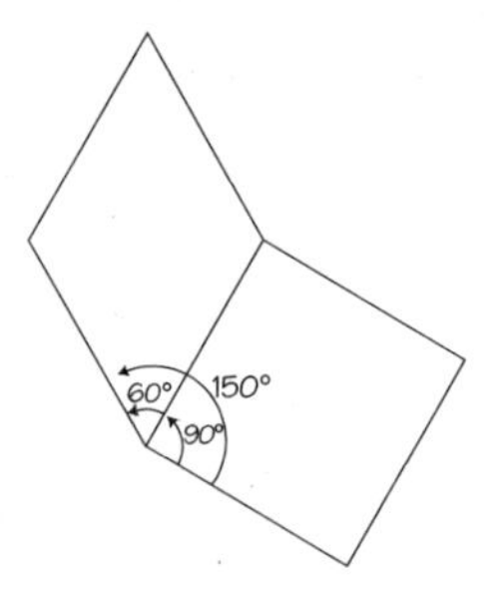

- Try to build more than one arrangement of blocks for each angle measure you find.
- Be ready to explain how you know you found all possible angles.

Part 2

What if... ***the mosaic is to contain polygons that have at least one angle of every possible measure? What polygons could the students use?***

- Using your work from Part 1, work with your partner to design at least three polygons for the mosaic. Your polygons may be any size and shape, but they must meet the following conditions:
  - Each polygon must contain at least one angle of every possible angle measure.
  - No polygon can contain more than three angles that have the same measure.
  - Each polygon must contain at least one tile of each of the six different Pattern Block shapes.
- Trace each of your polygons and the blocks used to make them onto white paper. Label the angle measures at each vertex.
- Be ready to tell how you went about creating the polygons for your mosaic.

Explain how you were able to do this activity without using a protractor. Use diagrams to help clarify your explanations.

What's Inside?

Part 1

What's Inside? is a game in which players use sets of Pattern Blocks to build polygons. The object of the game is to find the polygon whose interior angles have the smallest possible sum. Play What's Inside? following the rules below. Can you figure out the winning strategies?

- This is a game for 2 or 3 players. To begin, players take turns rolling a die 6 times, keeping track of the 6 numbers rolled. Players then assign each of their numbers to a *different* Pattern Block shape. Then players collect the assigned number of each block and place those blocks in front of them.
- For example, if a player rolls 6, 3, 2, 5, 5, 2, the player might collect:

6 yellow hexagons	3 orange squares
2 red trapezoids	5 green triangles
5 blue rhombuses	2 tan rhombuses

- Next, each player rolls the pair of dice once and adds that number of blocks to one opponent's set of blocks. The blocks do not have to be the same shape or color.
- Each player now uses all of his or her blocks to build a polygon. Blocks must be placed edge to edge, with no gaps or overlaps. The object is to try to arrange the blocks in such a way so that the sum of the interior angles of the completed polygon is as small as possible.

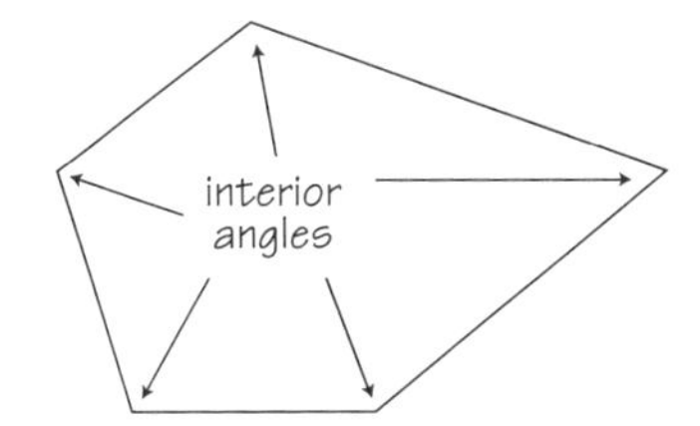

- Trace or sketch your final polygon. Label the measures of the interior angles on your drawing and record their sum.
- Play three rounds of *What's Inside?*. Add the sums from all three rounds. The player with the lowest combined total wins.

Part 2

What if... ***the rules of the game are changed a bit, and the object is now to make several polygons, each polygon having a different interior angle sum? What can you discover about the possible results of this version of the game?***

- Play *What's Inside?* with your group again, but this time, after making a polygon with the smallest possible sum, see what other sums you can obtain using the same set of blocks. Each player should try to make at least four or five polygons with different interior angle sums.
- Trace or sketch each of your polygons. Label the measures of the interior angles and record their sum.
- Compare all of the polygons recorded by the players in your group. See what you can conclude about possible interior angle sums and how they are related to the polygons that produce them.
- Be ready to discuss your findings.

Suppose you are in charge of creating a new set of Pattern Block shapes. Each new shape is to have congruent sides and congruent angles. Describe how you would designate the measure of each interior angle of each block.

Interior/Exterior

Part 1

Tai and Will have decided to make a book of Tangram puzzles. While drawing and measuring some of their puzzle outlines, they noticed something interesting about the angles of puzzles that have the same number of sides. Can you figure out what they discovered?

- Work with a partner. Using any combination from one set of Tangram pieces, make a shape that has four sides. Make sure your shape is different from your partner's shape. Record the outline of your shape.
- Using what you know about Tangrams, find and label the measure of each interior angle of your shape. Then record the sum of your angle measures.
- Make five more sets of shapes; each set should have a different number of sides. Record your findings as you did for the four-sided shapes.

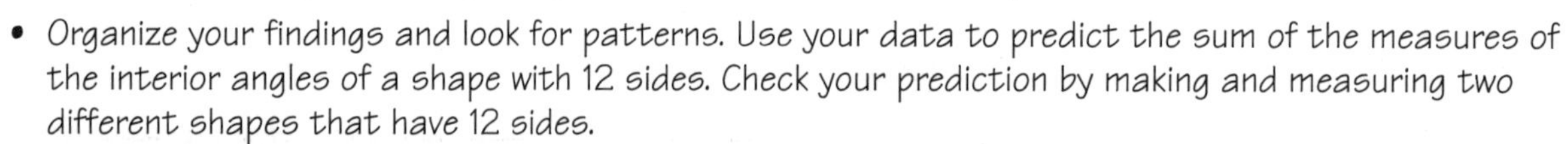

- Organize your findings and look for patterns. Use your data to predict the sum of the measures of the interior angles of a shape with 12 sides. Check your prediction by making and measuring two different shapes that have 12 sides.
- Be ready to explain your findings and to show how you can figure out the sum of the measures of the interior angles of any polygon with a given number of sides.

Part 2

What if... ***you measured the exterior angles of your shapes? Can you find a relationship between the sum of their measures and the number of sides in the shape?***

- Work with your partner. For each *convex* shape you recorded in the first activity, use a ruler to extend one of the sides at each vertex to form exterior angles. (If all or most of your shapes are concave, you may need to build some new shapes that are convex.)
- Determine and record the measure of each exterior angle of your shapes. Also record the number of sides of each shape and the sum of its exterior angle measures. Count only one exterior angle at each vertex.
- Organize your findings and look for patterns. Use your data to predict the sum of the exterior angles of a shape with 12 sides. Check your prediction as you did in Part 1.
- Be ready to discuss your findings about the sum of the measures of the exterior angles of a polygon.

Suppose you know the sum of the interior angles of a particular *regular* polygon (a polygon that is both equilateral and equiangular). Explain how you might figure out the number of sides in the polygon and the measure of each interior and each exterior angle.

Color Tile Grid Paper

1-Centimeter Grid Paper

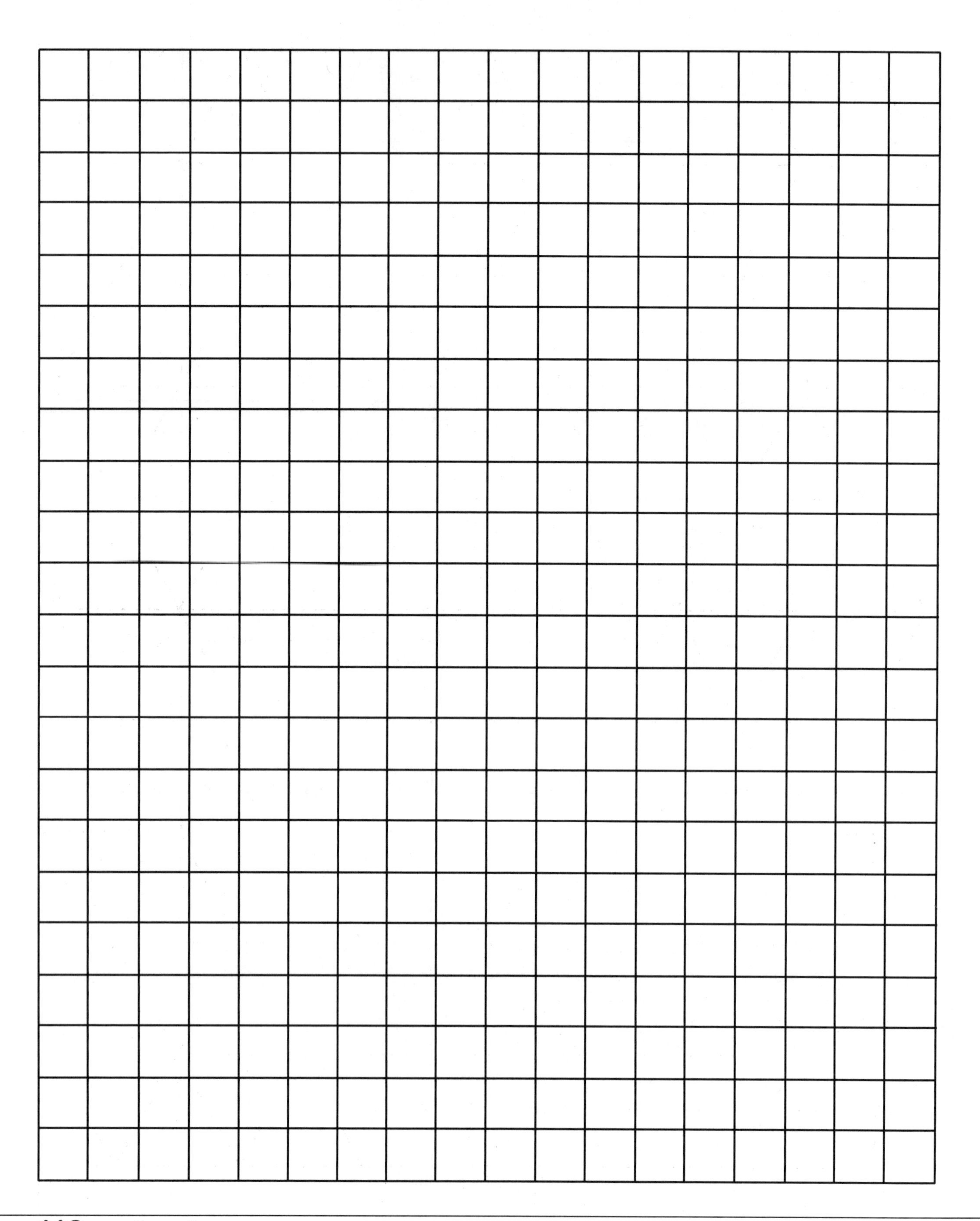

Pattern Block Triangle Paper

Isometric Dot Paper

Geodot Paper – 1 Grid

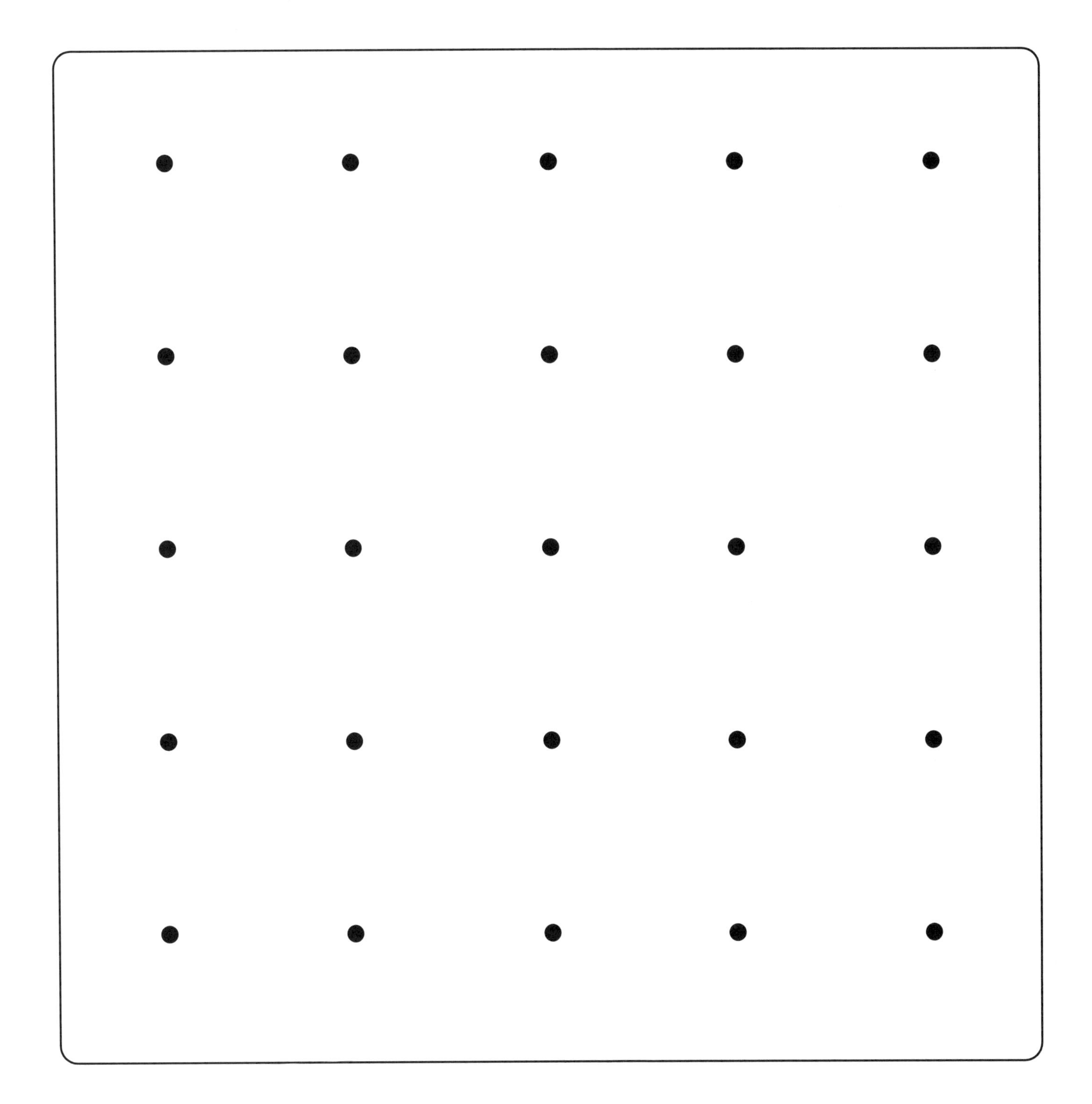

Tangram Paper

Hexagon Silhouette

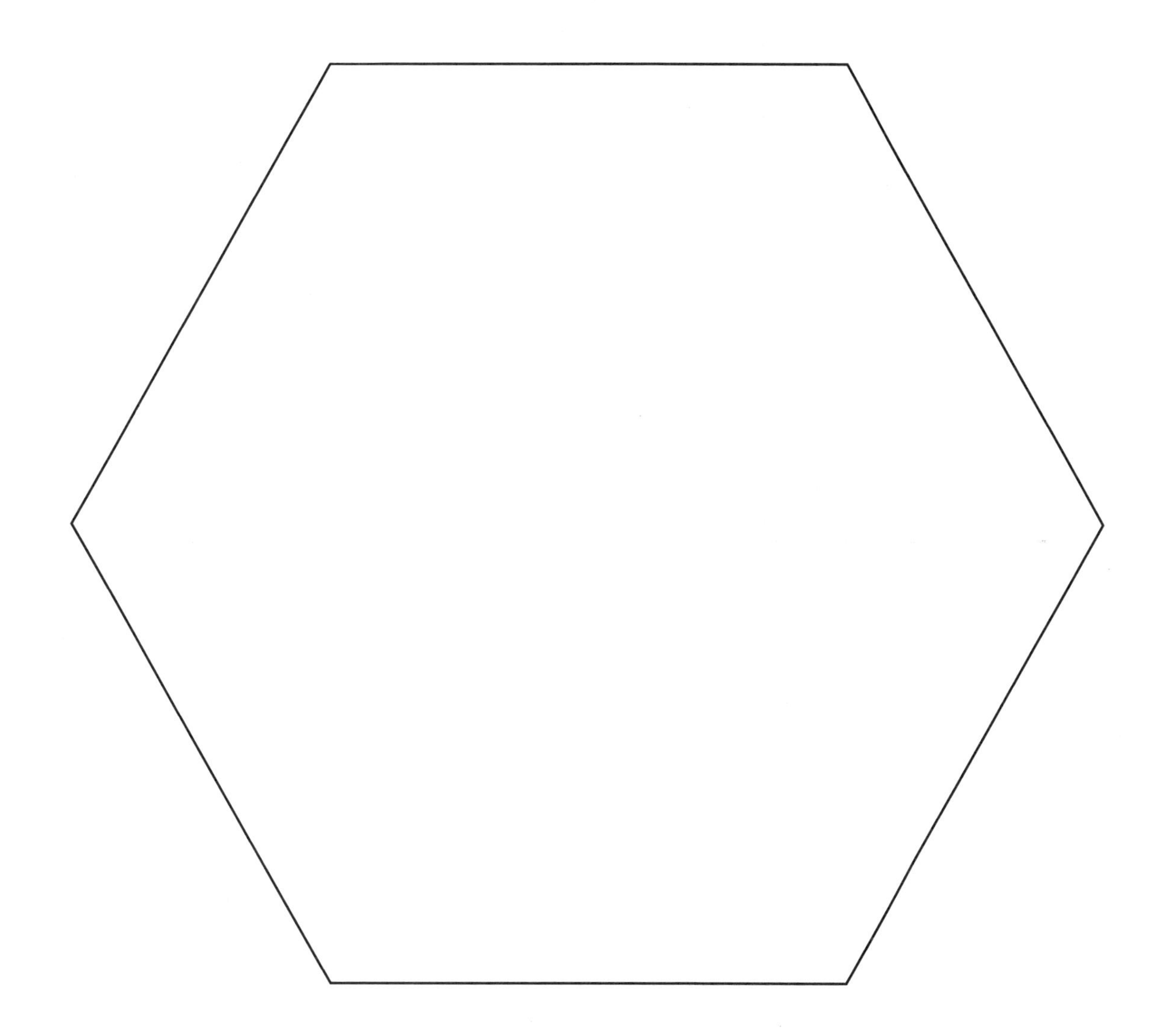

Snap Cube Grid Paper

Geodot Paper - 9 Grids

Circular Geodot Paper

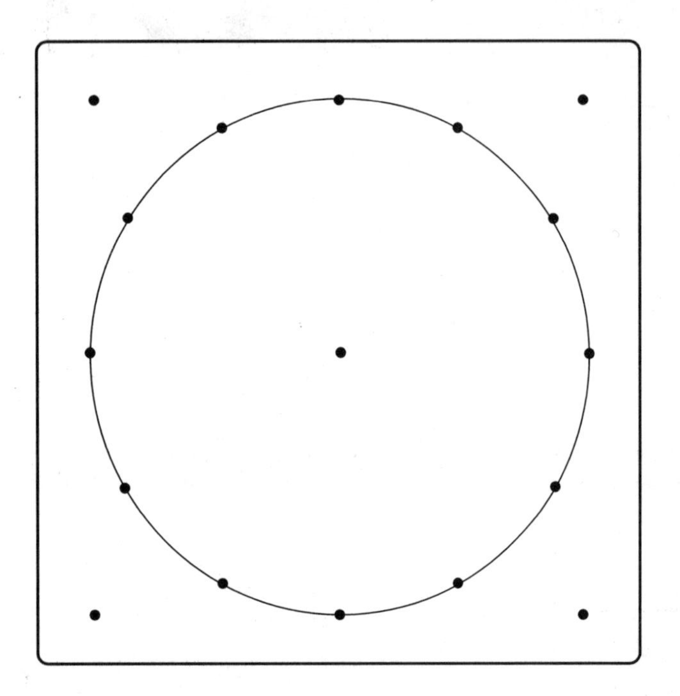
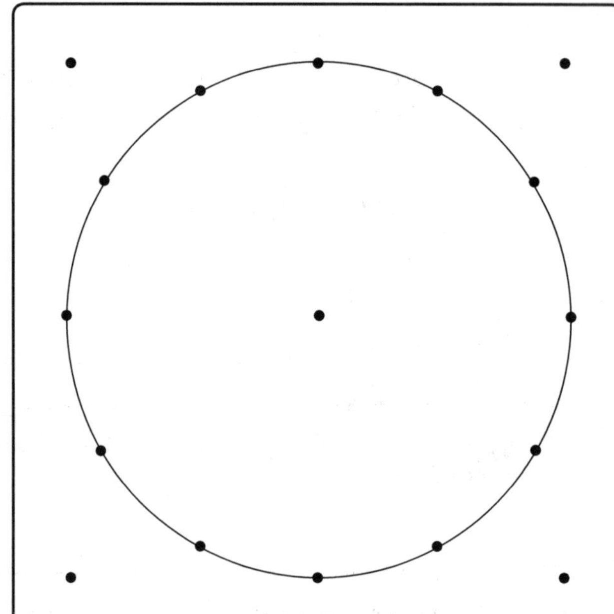
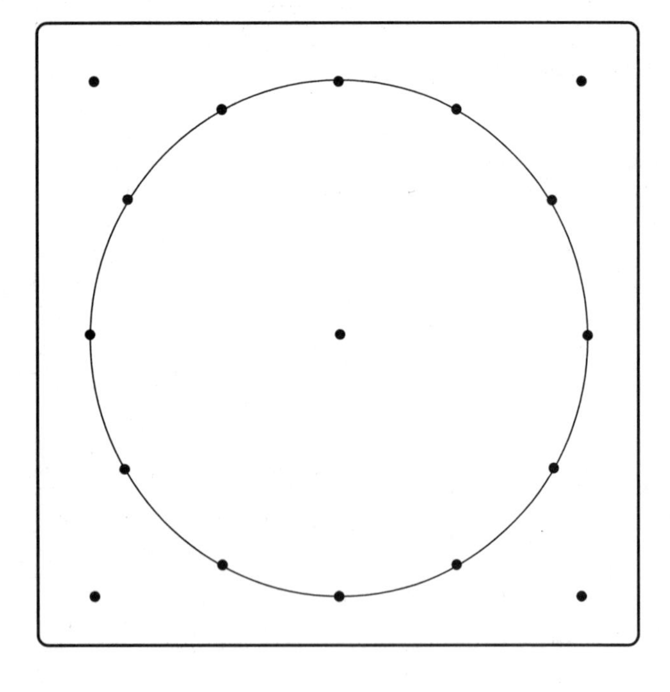
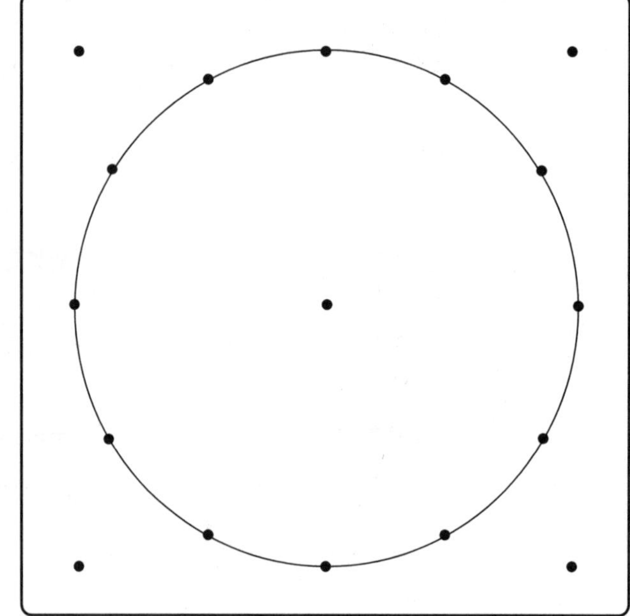

